# SCIENCE 1102

# ELEMENTS, COMPOUNDS, AND MIXTURES



INTRODUCTION |3

1. **ELEMENTS** 5

ALCHEMY |6
SYMBOLS |7
PROPERTIES |8
SELF TEST 1 |12

2. **COMPOUNDS** 14

PROPERTIES |15
CLASSIFICATION |28
SELF TEST 2 |30

3. **MIXTURES** 33

HETEROGENEOUS |34
HOMOGENEOUS |40
SELF TEST 3 |45

**PERIODIC TABLE** 50

**GLOSSARY** 52

**LIFEPAC Test is located in the center of the booklet.** Please remove before starting the unit.

**Author:**
Harold Wengert, Ed.D.

**Editor:**
Alan Christopherson, M.S.

**Media Credits:**
**Page 6:** © UGChannel, iStock, Thinkstock; **8:** © siambizkit, iStock, Thinkstock; **15:** © Cipariss, iStock, Thinkstock; **28:** © Diana Taliun, iStock, Thinkstock; **29:** © stephenallen75, iStock, Thinkstock; **50, 51:** NIST.

Alpha Omega PUBLICATIONS

**804 N. 2nd Ave. E.**
**Rock Rapids, IA 51246-1759**

# Elements, Compounds, and Mixtures

## Introduction

Chemistry is an ancient science. Early in Genesis (Genesis 4:22) record is made of man's use of chemistry. In man's quest for ways to "subdue the earth and have dominion over it," chemistry has played a major role. Our synthetic world of today with its many different forms of plastic is a result of the knowledge man has gained about chemistry.

On the other hand, our polluted streams, dirty atmosphere, and trash-laden landfills are also a result of our knowledge of chemistry. Knowledge is neither good nor bad, but man's use of it can be either good or bad. Because of man's innate sinful nature, self is first, encouraging the misuse and poor stewardship of God's creation.

In this LIFEPAC® you will study the history of chemistry, the classification of matter, and man's knowledge of the world about him.

## Objectives

**Read these objectives.** The objectives tell you what you will be able to do when you have successfully completed this LIFEPAC. When you have finished this LIFEPAC, you should be able to:

1. Trace the history of chemistry to about 1750.
2. Describe alchemy and what it has contributed to us.
3. Describe elements.
4. Distinguish between and describe chemical, physical, and phase changes.
5. Describe compounds and distinguish them from elements.
6. Classify compounds as organic or inorganic.
7. Describe and give examples of various types of mixtures.

Survey the LIFEPAC. Ask yourself some questions about this study and write your questions here.

# 1. ELEMENTS

All matter can be classified in the following classification scheme:

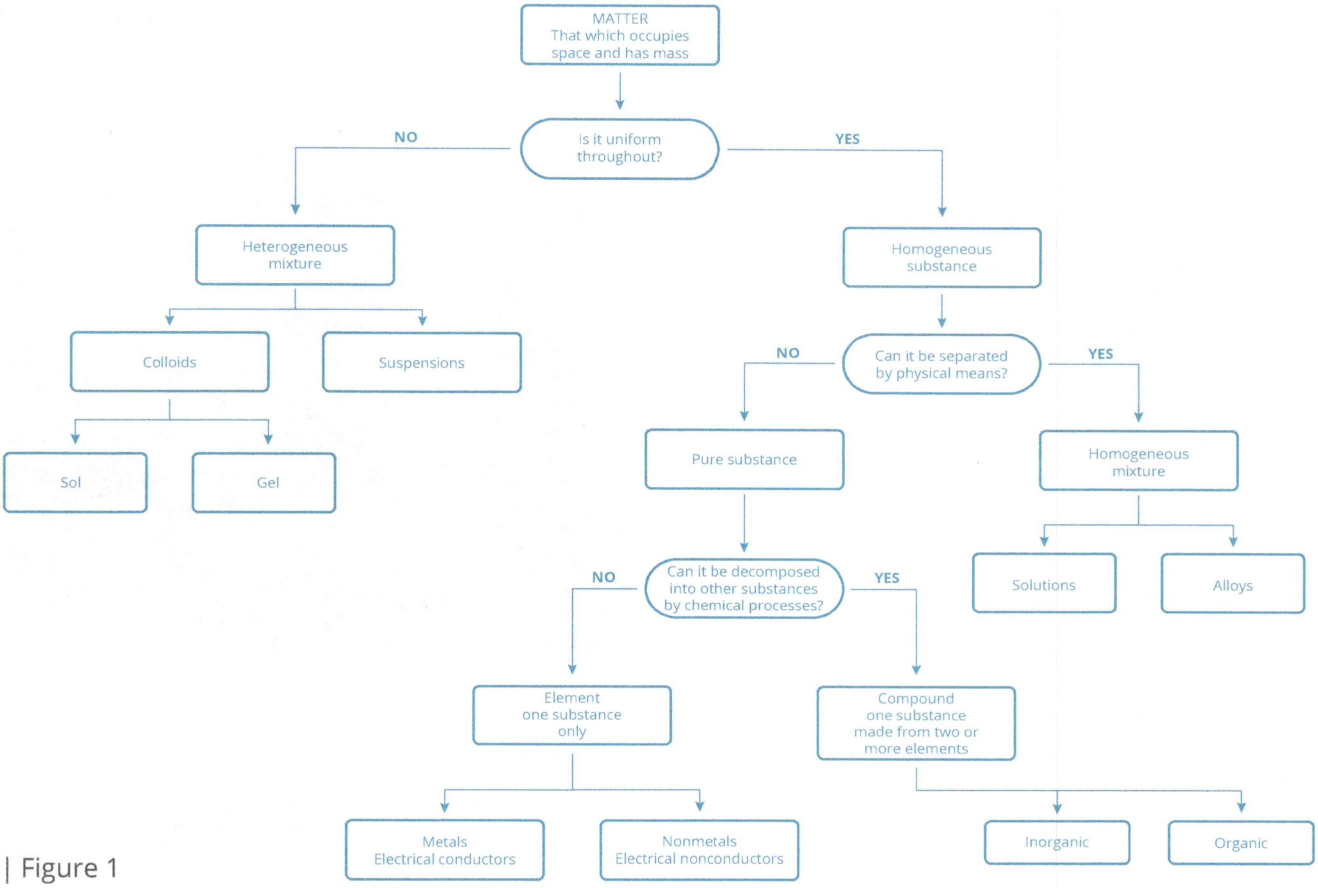


| Figure 1

We will use this classification scheme for our study of chemistry.

In man's effort to find order in the universe, he has learned to classify. Man has always been curious about the world around him. The word *science* comes to us from the Latin verb *scio* which means *to know*. The Latin noun *scientia* means knowledge. In both cases the meaning is broader than mere recognition. In this section you will study the history of alchemy, the contributions of alchemy, and the properties of the simplest parts of matter, elements.

## Section Objectives

**Review these objectives.** When you have completed this section, you should be able to:

1. Trace the history of chemistry to about 1750.
2. Describe alchemy and what it has contributed to us.
3. Describe elements.
    3.1 Give the symbols of several common elements.
    3.2 Classify elements by properties.
    3.3 Define atoms.

## Vocabulary

**Study these words to enhance your learning success in this section.**

alchemy alloy atom compound element metal nonmetal

**Note:** *All vocabulary words in this LIFEPAC appear in* **boldface** *print the first time they are used. If you are not sure of the meaning when you are reading, study the definitions given.*

# ALCHEMY

Who was the first chemist? No one really knows, but Genesis 4:22 clearly describes Tubalcain as an art teacher who worked with brass and iron. This information indicates that chemistry had been established as a science by then. However, that chemistry was probably limited to simple metallurgy and the use of metals.

| Alchemy tools

**History.** As civilization grew, the center of all scientific knowledge became Alexandria, Egypt. About 300 B.C. the science of **alchemy** became important. The Egyptians were interested in making life easier. As a result, a deliberate effort was made to combine the Egyptian arts of metallurgy, dyeing, and glassmaking with a search for a "better" life. As a result, the alchemist tried to:

1. Discover a universal cure for disease,
2. Discover a means to prolong life, and
3. Transmute the base elements like lead into gold.

**Prepare a report.**

**1.1** The history of the spread of alchemy to Europe from 300 B.C. through the Dark Ages and up to about 1750 is interesting because it was greatly influenced by the religious beliefs of the nations in control. Aristotle, Descartes, and Roger Bacon are examples of people who influenced the growth of chemistry throughout this period. Find out what "puffers" and "adepts" were. Prepare a report on the growth and spread of alchemy from 300 B.C. to about A.D. 1750. Include names, dates, and specific influences the church had on this spread. Submit the report for evaluation.

**CHECK** ______________ ______________
Teacher Date

**Legacy.** Early alchemists contributed much to modern knowledge of matter. They were able to isolate and study the properties of common **elements** such as copper, sulfur, iron, lead, gold, silver, arsenic, and mercury. They also studied compounds such as mercury oxide and **alloys** of **metals** and developed laboratory techniques that were used by later scientists.

The transition from alchemy to the science of chemistry occurred when the basis of the study became truly experimental in practice. Scientists in the sixteenth century began to put each theory to the test and found the laboratory the heart of the learning process.

**Complete these statements.**

**1.2** The two main classes of homogeneous matter are a. ________________ and b. ________________ .

**1.3** Pure substances can be classed as a. ________________ or b. ________________ .

**1.4** The three purposes of alchemy were:

a. ________________________________________ ,

b. ________________________________________ , and

c. ________________________________________ .

**1.5** The city that was first known for alchemy was ________________________ .

## SYMBOLS

Very early, even in the days of alchemy, chemists discovered the need for symbols to express what they were doing. Using symbols provided a simple way to communicate without always writing out the full words or names.

**Derivation.** Early symbols were derived from the surroundings, things that were simple and easily understood.

**Representation.** Today's symbol representation is derived primarily from an abbreviation of the name of the element, either the English or Latin spelling. A list of common elements and symbols is found in Figure 3. These should be memorized.

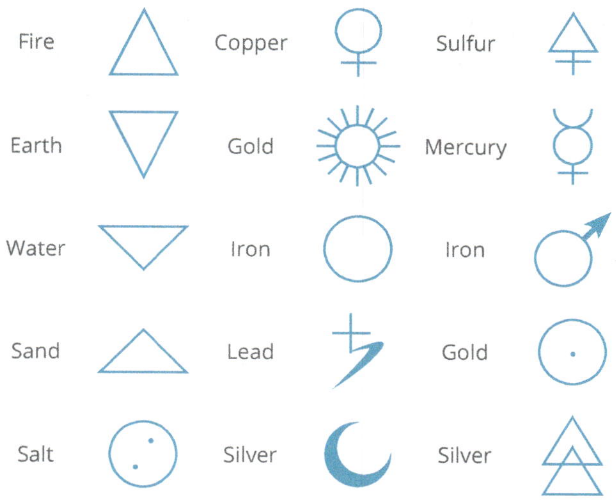


| Figure 2: Early symbols were derived from surroundings.

| Name | Symbol | Name | Symbol | Name | Symbol |
|---|---|---|---|---|---|
| Hydrogen | **H** | Silver | **Ag** | Aluminum | **Al** |
| Carbon | **C** | Lead | **Pb** | Zinc | **Zn** |
| Nitrogen | **N** | Nickel | **Ni** | Gold | **Au** |
| Calcium | **Ca** | Oxygen | **O** | Magnesium | **Mg** |
| Copper | **Cu** | Neon | **Ne** | Uranium | **U** |
| Iron | **Fe** | Helium | **He** | Tin | **Sn** |
| Mercury | **Hg** | Iodine | **I** | Arsenic | **As** |

| Figure 3: Names and symbols of common elements

**Answer these questions.**

**1.6** What is common to all first letters of the element symbol? ______________________

______________________

**1.7** When two names begin with the same letter (i.e., hydrogen and helium), how are the two represented so that they can be distinguished from each other?

______________________

______________________

**1.8** Notice that several elements have symbols quite different from their names. Do these symbols come from the English or Latin name? ______________________

**Complete this activity.**

**1.9** Fill in the following table.

| Symbol | Name of Element |
|---|---|
| Cu | |
| Pb | |
| Hg | |
| Ag | |
| Au | |

## PROPERTIES

Each element has its own unique properties. Those properties identify it as that element, and no other element has exactly those same characteristics.

**Physical.** Some of the easiest ways to identify a particular element is by its physical properties.

These properties include such things as color, hardness, boiling point, melting point, density, magnetic properties, crystalline shape, and ability to conduct electricity.

| What are some characteristics of this sulfur?

**Try this investigation.**

**These supplies are needed:**

- flashlight battery with light bulb, socket, and wires
- sample of aluminum, zinc, carbon, iron, copper, and sulfur

**Follow these directions and complete the activities. Put a check in the box when each step is completed.**

☐ 1. Put together the conductivity tester as shown. The ends of the wire can be taped to the contacts of the battery with electrical tape.

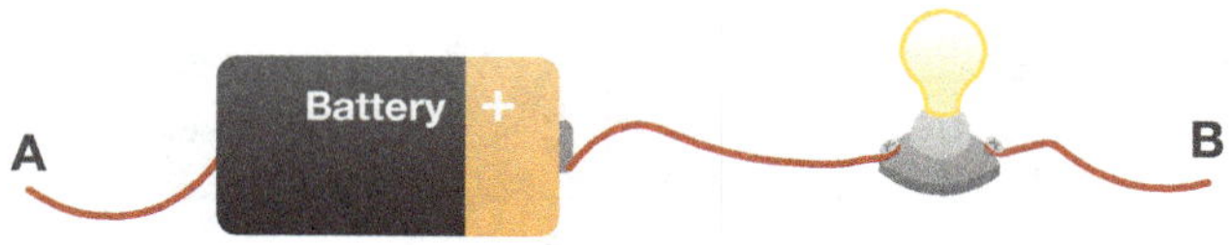


☐ 2. Touch the two loose ends, A and B, together to see if the bulb lights. If it doesn't light, recheck your circuit. If it still doesn't light, change bulbs and/or batteries. Be sure the bulb does light before you go on.

☐ 3. Using each sample separately, touch A to one end and B to the other end of the sample. For sulfur, if it is in powder form, place the points A and B in the powder about 1 cm apart. Record your results in the second and third columns of the chart in 1.10.

## Conductivity Experiment

**Complete these items.**

**1.10** Give the results of your investigation. Complete columns two and three.

| Element | Symbol | Conducts (yes / no) | Classification |
|---|---|---|---|
| Aluminum | | | |
| Zinc | | | |
| Carbon | | | |
| Iron | | | |
| Copper | | | |
| Sulfur | | | |

**1.11** Look back at the classification scheme for matter in Figure 1. On the basis of the classification of elements, classify the test elements as metals or **nonmetals**. Complete the last column of the chart in 1.10.

**Chemical.** All elements can be classified by their chemical properties also. This classification is more difficult, however, because elements tend to have similar chemical properties and can be grouped by these properties.

The smallest unit that can exist as an element and still have the properties of that element is called an **atom**. All atoms of the same element have the same chemical properties. For example, all atoms of copper behave the same when reacted with oxygen. This regularity illustrates clearly the master plan and orderliness of the Creator. If random chance were in control, copper atoms would be produced randomly and would not all behave the same way because they would not necessarily be the same.

Over the years scientists have classified all the elements chemically and arranged them in a chart called a Periodic Table. Some Periodic Tables include elements with atomic numbers up to 118. These tables extend beyond the approximately 98 naturally occurring elements to those that have been or are thought possible to be synthesized. Study the following Periodic Table to see the location of the different elements.

One of the reasons for the present arrangement of the Periodic Table is that results of repeated experimentation have shown that elements in the same column have similar chemical properties. For example, if you know that H reacts with O to form $H_2O$, you would expect Li to react with O to form $Li_2O$.

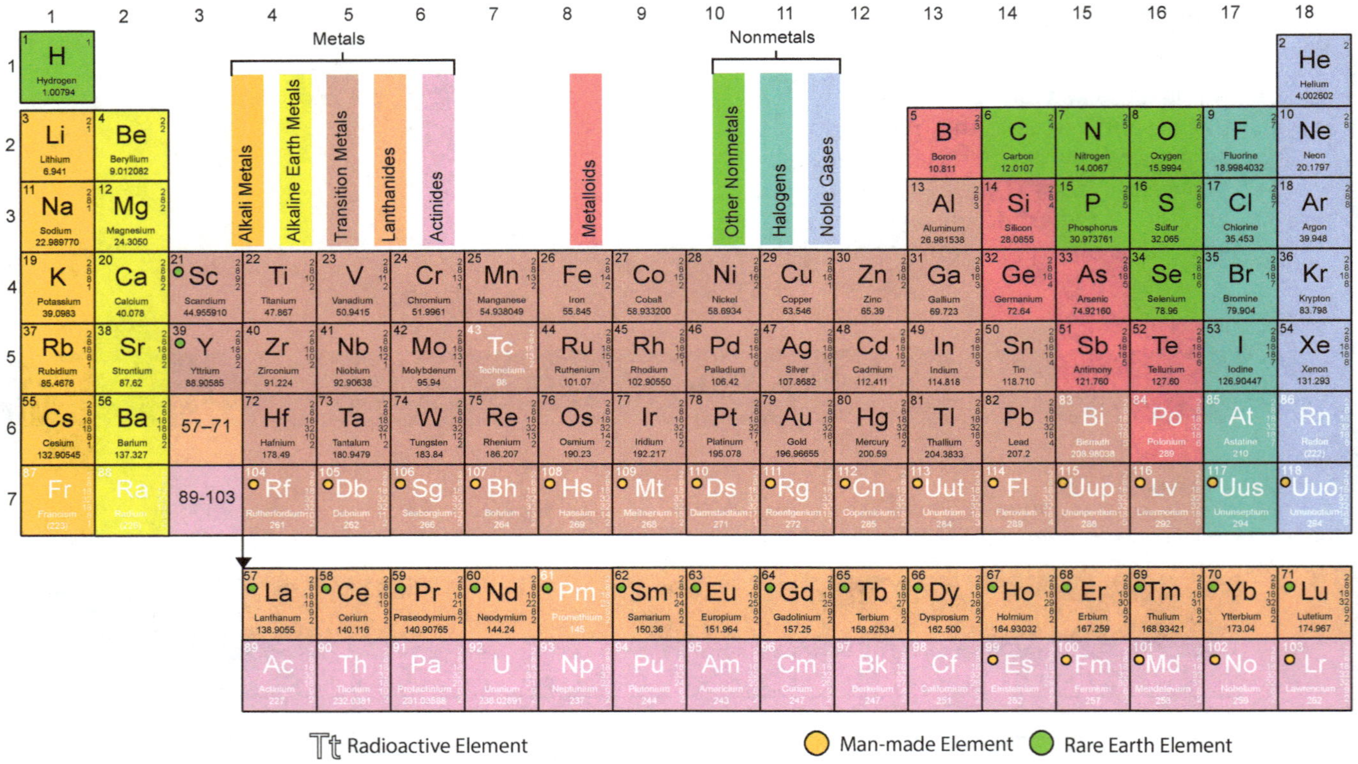


| Figure 4: Periodic Table

**Complete these activities.**

**1.12** Apply the preceding idea about the Periodic Table and complete the following chart.

| Formula | New Combination | Predicted Formula |
|---|---|---|
| $Cu_2O$ | silver + oxygen | |
| $NaCl$ | potassium + chlorine | |
| $AlCl_3$ | aluminum + fluorine | |
| $CO_2$ | tin + oxygen | |
| $MgCl_2$ | calcium + bromine | |
| $HCl$ | cesium + iodine | |
| $CCl_4$ | silicon + bromine | |

**1.13** Based on the Periodic Table, answer these questions about the experiment in 1.10.

a. Was your data correct for the experiment? ____________________

b. If not, what was wrong? ____________________

____________________

**1.14** Use the Periodic Table to answer these questions.

a. How many elements are metals? ____________________

b. How many elements are nonmetals? ____________________

**1.15** Describe the location of the metals and nonmetals on the Periodic Table.

____________________

____________________

____________________

**CHECK** ____________ ____________

Teacher Date

**Review the material in this section in preparation for the Self Test.** The Self Test will check your mastery of this particular section. The items missed on this Self Test will indicate specific areas where restudy is needed for mastery.

# SELF TEST 1

**Match these items** (each answer, 2 points).

| | | | |
|---|---|---|---|
| 1.01 | ________ | hydrogen | a. O |
| 1.02 | ________ | copper | b. Pb |
| 1.03 | ________ | zinc | c. Al |
| 1.04 | ________ | gold | d. C |
| 1.05 | ________ | silver | e. Ag |
| 1.06 | ________ | oxygen | f. H |
| 1.07 | ________ | nitrogen | g. Cu |
| 1.08 | ________ | carbon | h. Zn |
| 1.09 | ________ | aluminum | i. Au |
| 1.010 | ________ | lead | j. N |
| | | | k. Le |
| | | | l. Zi |

**Complete the following statements** (each lettered item, 3 points).

**1.011** The two main classes of homogeneous matter are a. ______________ and b. ______________ .

**1.012** A pure substance can be grouped as a. ______________ or as b. ______________ .

**1.013** An alchemist had three goals. They were:

a. ________________________________________ ,

b. ________________________________________ , and

c. ________________________________________ .

**1.014** The modern symbols for the elements are derived from either the a. ______________ or b. ______________ spelling.

**1.015** Elements are classed as a. ______________ or b. ______________ .

**Based on the Periodic Table** (Figure 4), complete the following chart (each answer, 3 points).

| | Formula | New Combination | Predicted Formula |
|---|---|---|---|
| 1.016 | $NaCl$ | sodium + bromine | |
| 1.017 | $MgO$ | calcium + oxygen | |
| 1.018 | $K_2O$ | cesium + sulfur | |
| 1.019 | $CaI_2$ | magnesium + bromine | |

| | Formula | New Combination | Predicted Formula |
|---|---|---|---|
| 1.020 | $H_2O$ | potassium + sulfur | |
| 1.021 | $Al_2O_3$ | boron + sulfur | |
| 1.022 | $Ag_2O$ | gold + sulfur | |

**Match these items** (each answer, 2 points).

1.023 ________ smallest unit of an element
1.024 ________ puffer
1.025 ________ original site of alchemy
1.026 ________ metal
1.027 ________ only one substance
1.028 ________ nonmetal
1.029 ________ adept
1.030 ________ greatly influenced the spread of alchemy

a. professional alchemist
b. carbon
c. element
d. iron
e. amateur alchemist
f. Babylon
g. molecule
h. atom
i. Alexandria
j. church

**Answer these questions using complete sentences** (each answer, 5 points).

1.031 Briefly explain the major points of history and the spread of alchemy from about 300 B.C. to A.D. 1750. ____________________________________________

____________________________________________

____________________________________________

____________________________________________

____________________________________________

1.032 What is alchemy and what has it contributed to modern chemistry?

____________________________________________

____________________________________________

____________________________________________

____________________________________________

____________________________________________

____________________________________________

80/100 **SCORE** ____________ **CHECK** ____________ ____________
Teacher Date

# 2. COMPOUNDS

As you found in Section 1 of this LIFEPAC, elements are pure substances that are composed of just one kind of atom. Naturally, elements are not found in their pure state very often. Some examples are gold, silver, and carbon. More frequently, the elements are found in combination of two or more together in the form of compounds. In this section of the LIFEPAC, you will study how pure compounds are formed, what their characteristics are, and how they can be classified.

## Section Objectives

**Review these objectives.** When you have completed this section, you should be able to:

4. Distinguish between and describe chemical, physical, and phase changes.
   - 4.1 Explain kinetic energy.
   - 4.2 Explain potential energy.
   - 4.3 Interpret the various parts of a phase change curve.
   - 4.4 Define reaction and products.
5. Describe compounds and distinguish them from elements.
   - 5.1 Describe the Law of Constant Composition.
   - 5.2 Classify by formula as compound or element.
6. Classify compounds as organic or inorganic.

## Vocabulary

**Study these words to enhance your learning success in this section.**

**chemical change** **physical change** **phase change**

**organic** **inorganic**

Compounds are defined as two or more elements combined with each other such that:

1. The elements can no longer be identified by their original individual properties and
2. Only a chemical reaction can tear them apart again.

## PROPERTIES

The properties of compounds are different from the properties of the elements that make them up. For example, the elements of hydrogen and oxygen (two gases) when reacted together produce common, everyday water. Sodium, a shiny metal that reacts violently with water, and chlorine, a pale green poisonous gas, react together to form harmless, water-soluble table salt.

**Physical.** The physical properties of compounds are much the same as those for elements. Compounds have specific densities, melting points, boiling points, crystalline shape, and solubilities in liquids. The elements making up the compound cannot be separated from each other by physical means, and each element has given up its individual identity to be replaced by the identity of the compound.

Science LIFEPAC 1101 gave you an opportunity to improve your skills of observing and hypothesizing. LIFEPAC 1102 provides you opportunities to use these improved skills to study the nature of matter. Matter comes in three normal physical states called phases. These states of matter are called solid, liquid, or gas. The basic difference between the solid, liquid, or gaseous state of any substance is the amount of energy the substance contains. Energy change in a system can occur in many ways and, depending upon the system, will display varying characteristics. LIFEPAC 1102 will help you to study the energy changes in matter. Everyone has some ideas about energy.

| Sucrose (table sugar) is a compound. Its formula is $C_{12}H_{22}O_{11}$.

---

**On the following lines, write the ideas that you have at this time about energy.**

**2.1** ______________________________________________

______________________________________________

______________________________________________

**You will now investigate one physical property of compounds.**

**These supplies are needed:**

- test tube with 12-15 grams paradichlorobenzene (PDCB)
- two 250 ml Pyrex beakers
- beaker stand
- heat source
- two thermometers
- water
- dye tablet or egg-coloring tablet

**Before investigating the physical properties of a compound, do this activity.**

**2.2** You will perform a three-part study of a simple system which will present some ideas about the interaction of energy with matter.

a. Fill a beaker about three-fourths full of tap water. When the water is no longer swirling (one or two minutes), drop one tablet of dye or egg coloring in the center of the beaker. Observe the system for a short period of time. Record your observations.

______________________________________________

______________________________________________

______________________________________________

______________________________________________

b. Set the beaker aside gently, label properly, and observe after it has stood for twenty hours. While this experiment is going on, continue with Activities 2.3 and following. Return to activity 2.2 to answer b and c after twenty-four hours have elapsed. Observations after twenty-four hours:

______________________________________________

______________________________________________

______________________________________________

c. Why do you think the change occurred as it did after twenty-four hours?

______________________________________________

______________________________________________

______________________________________________

**Physical Properties Experiment** (Continued on next page)

**Try another mixing activity.**

**2.3** Get two Pyrex beakers and fill one beaker half full of ice water and the second half full of boiling water. When the water in both beakers has stopped swirling, put one tablet in each. What do you observe?

______________________________________________

______________________________________________

______________________________________________

______________________________________________

**Investigate kinetic energy and answer the questions.**

**2.4** *Kinetic energy is the energy of motion.* The more motion an object has, the more energy it contains. For example, the more energy you give a bowling ball, the faster it goes. In other words, if the speed of a particle is increased, its kinetic energy is increased.

a. Does hot water or cold water have more energy? ____________________

b. Are the water particles moving faster in the hot or cold water? ____________________

c. As the temperature of a substance increases, does its speed of motion increase or decrease? ____________________

d. How is temperature related to speed of the particles of a substance? ____________________

______________________________________________

______________________________________________

**2.5** One way to study energy is to study the melting and cooling characteristics of a pure substance.

What is a *pure* substance? ______________________________________________

______________________________________________

______________________________________________

______________________________________________

**2.6** Using the ideas of kinetic energy given, hypothesize as to why you observed what you did in 2.3. Be specific.

______________________________________________

______________________________________________

______________________________________________

______________________________________________

**Physical Properties Experiment** (Continued on next page)

**Do these investigations.**

Carefully read this entire activity before you do the experiment. **Be sure to wear goggles.** Put a check mark in the box when each step is completed.

☐ 1. Put about 200 ml of tap water in a 250 ml Pyrex beaker.

☐ 2. Heat this system to a temperature of 75°-80° C.

☐ 3. While the water is heating, obtain a thermometer and a test tube containing solid paradichlorobenzene (PDCB) with a second thermometer hardened in the solid. (Another name for PDCB is moth balls or moth crystals.) Note: To prepare the thermometer hardened in the PDCB, add the PDCB to an empty test tube, melt the solid by putting the test tube with PDCB into a water bath of boiling water, put a thermometer in the melted PDCB, and allow to cool with the thermometer in place.

☐ 4. When the water bath has reached 75°-80° C, *remove the heat source*.

☐ 5. Clamp the test tube of PDCB in the hot water bath and immediately read the temperature of both the PDCB and hot water bath.

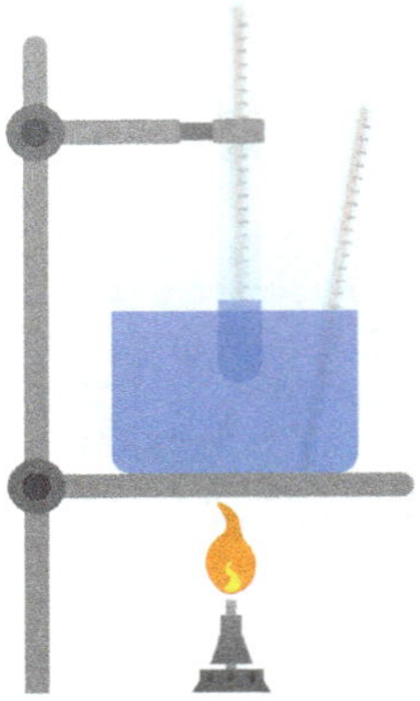

☐ 6. In the following chart, record the temperatures every 30 seconds until the two temperatures are within a degree of each other. As the PDCB melts, gently stir the solid by slowly moving the thermometer up and down. This procedure will give more accurate temperatures. Also record the observations at temperatures where significant changes in the properties of the PDCB occur. Keep the PDCB for activity 2.8.

2.7

**Data Table 1**

| Elapsed Time | Temperature | | Observations | Elapsed Time | Temperature | | Observations |
|---|---|---|---|---|---|---|---|
| min:sec | PDCB | Water | | min:sec | PDCB | Water | |
| 0:00 | | | | 4:30 | | | |
| 0:30 | | | | 5:00 | | | |
| 1:00 | | | | 5:30 | | | |
| 1:30 | | | | 6:00 | | | |
| 2:00 | | | | 6:30 | | | |
| 2:30 | | | | 7:00 | | | |
| 3:00 | | | | 7:30 | | | |
| 3:30 | | | | 8:00 | | | |
| 4:00 | | | | 8:30 | | | |

✓ **CHECK** ______________ ______________
Teacher Date

**Physical Properties Experiment**

(Continued on next page)

**WEAR GOGGLES**

Carefully read this whole activity before you do the experiment. In the previous questions, you heated PDCB in a hot water bath until it became a liquid. Now let's see what happens if we start with liquid PDCB and watch it cool.

Put a checkmark in the box when each step is completed.

- ☐ 1. Warm the PDCB in the test tube used in 2.7 in a hot water bath until the PDCB is at 75°–80° C.
- ☐ 2. While the PDCB in the test tube is warming, get another 250 ml beaker and place about 100 ml of 25°C tap water in it.
- ☐ 3. When the PDCB has reached the desired temperature, place the warmed PDCB in the beaker of tap water, as you did in 2.7.
- ☐ 4. Read the temperatures of both PDCB and water bath immediately. Record these temperatures in Data Table 2.
- ☐ 5. Continue recording temperatures every 30 seconds.
- ☐ 6. Record observations of significant changes in the physical nature of the PDCB. Stop recording when the two temperatures are within one degree of each other.
- ☐ 7. Why is the data beyond this point not significant?

______________________________________________

______________________________________________

______________________________________________

______________________________________________

______________________________________________

______________________________________________

**Data Table 2**

| Elapsed Time min:sec | Temperature PDCB | Temperature Water | Observations | Elapsed Time min:sec | Temperature PDCB | Temperature Water | Observations |
|---|---|---|---|---|---|---|---|
| 0:00 | | | | 6:30 | | | |
| 0:30 | | | | 7:00 | | | |
| 1:00 | | | | 7:30 | | | |
| 1:30 | | | | 8:00 | | | |
| 2:00 | | | | 8:30 | | | |
| 2:30 | | | | 9:00 | | | |
| 3:00 | | | | 9:30 | | | |
| 3:30 | | | | 10:00 | | | |
| 4:00 | | | | 10:30 | | | |
| 4:30 | | | | 11:00 | | | |
| 5:00 | | | | 11:30 | | | |
| 5:30 | | | | 12:00 | | | |
| 6:00 | | | | 12:30 | | | |

**Physical Properties Experiment**

**Complete these graphing activities.**

**2.9** One of the best ways to handle numerical data like we collected in Data Tables 1 and 2 is to graph the data. Graphing gives us a visual picture of what has happened and is very useful in observing important changes. On one graph plot a smooth curve for the two sets of data from Data Table 1. Plot the temperature on the *y*-axis (vertical axis) and the time on the *x*-axis (horizontal axis). Label the graph completely.

a. Have you labeled your graph? ______________

b. Does it have units on each axis? ______________

c. Does it have a title? ______________

d. What does your graph represent? ______________

______________

______________

______________

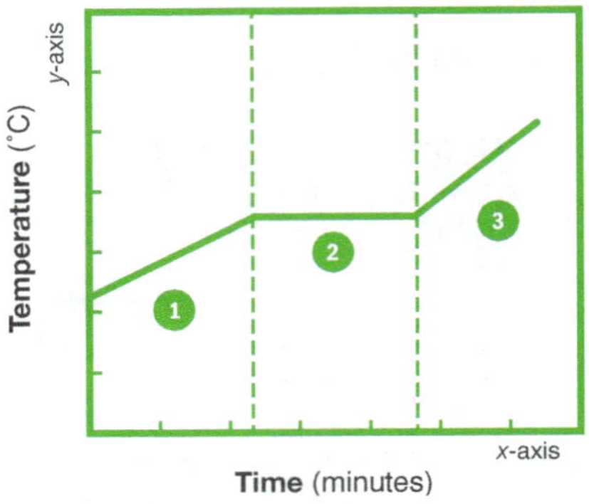


| Figure 5

**2.10** Curves that graph experimental data can be divided into distinct parts based on a change in slope. For example, the curve shown in Figure 5 has three distinct parts. Look closely at your graph of the warming curve from 2.7. Your plot should be similar in shape to the example in Figure 5. Label your graph in a similar way by dividing it in three parts such as the example.

a. According to your data and observations from 2.7, what state of matter is PDCB in during warming as in Part 1 of the example in Figure 5? (Solid, liquid, or gas)

b. What is happening in Part 2? (freezing, melting, turning into a gas)

c. Generally, what state of matter is PDCB in for Part 3? (Solid, liquid, gas)

d. Based on this data, what do you think is the melting point of PDCB? The accepted value for the melting point of paradichlorobenzene is 53.1°C.

**Answer these questions.**

**2.11** From Data Table 1, how can you tell that PDCB is gaining energy? ______________

______________

______________

______________

**2.12** Where is the energy coming from to warm the PDCB? a. ______________ How can you tell?

b. ______________

______________

**2.13** Look again at your graph of the warming (2.10) of PDCB. What two parts of the curve represent an increase in the kinetic energy of the PDCB particles?

a. ______________ and b. ______________

c. How can you tell? ______________

______________

**2.14** Something else must be happening in Part 2 of the data because the temperature is not changing, yet energy is being added continuously to the PDCB test tube. How do you know? (Refer to 2.15)

______________________________________________

______________________________________________

**Investigate potential energy and answer the questions.**

**2.15** When energy is gained by a system but no change is evident in the temperature, the energy that is gained is stored as potential energy. Potential energy is used to break the particles apart from one another but not to change their speed of motion.

a. What part of your graph from the data in 2.7 shows an increase in potential energy?

______________

b. How can you tell? ______________________________________________

______________________________________________

**2.16** What is the shape of the graph when there is a potential energy change but no change in kinetic energy?

______________________________________________

______________________________________________

✓ **CHECK** ______________ ______________
Teacher Date

**Make a graph; then answer the questions.**

**2.17** On a second piece of graph paper, plot the data from Data Table 2 using the same procedure and labels as you did in 2.9.

a. What are your units? ______________

b. What is its title? ______________________________

c. Divide the data into three parts and label in a way similar to what you did in 2.10.

d. What is the melting point (freezing point) of PDCB as determined by the cooling curve?

______________

e. Are the freezing point and melting point equal? ______________

f. Should they be equal? ______________

g. Why or why not? ______________________________________________

______________________________________________

______________________________________________

**Answer the questions using information from 2.17.**

**2.18** Look carefully at Part 1 of the data plotted in 2.17.

a. Does PDCB show an increase or decrease in energy? ____________________

b. What kind, kinetic or potential? ____________________

c. How can you tell? ____________________

____________________

**2.19** What is happening and why in Part 2 of the PDCB curve plotted in 2.17? Explain in detail, using the term "particles" in your answer to describe how the temperature is changing, how the potential and/or kinetic energy is changing, and how the phases are changing.

____________________

____________________

____________________

____________________

**2.20** What is happening to the PDCB and why as shown by Part 3 of the graph plotted in 2.17? Explain in detail; be sure to discuss what is happening to the particles and the type of energy change involved.

____________________

____________________

____________________

**Answer these questions.**

**2.21** If a substance were half melted (half liquid and half solid) and the surroundings were kept at exactly the same temperature as the melting point of the substance,

a. Would the substance melt, turn into a solid, or neither? ____________________

b. What is your reasoning? ____________________

____________________

____________________

____________________

**2.22** Does the liquid state of matter have more or less total energy than the same material as a solid?

a. ____________________

b. Therefore, does the liquid state have more or less kinetic energy than the same material as a solid?

____________________

**2.23** The name given to what we have studied with PDCB is **phase change**. Phase changes are a type of **physical change** that involve energy changes. Phase change can be represented as

Solid + Energy ⟶ liquid
$H_2O$ (s) + heat ⟶ $H_2O$ (l)     s = solid, l = liquid
Wax (s) + heat ⟶ Wax (l)

A physical change can also be just a change in the physical characteristics of a substance such as shape, size, or crystalline appearance. In these examples the phase stays the same, however:

a. Are all phase changes physical changes? ____________________

b. Are all physical changes phase changes? ____________________

c. Why or why not? ____________________________________________

________________________________________________________________

________________________________________________________________

✓ **CHECK** ____________________ ____________________
Teacher Date

**2.24** When the PDCB melted, was it still PDCB?

a. ____________________

b. When ice melts, is it still water? ____________________

c. When wax melts, is it still wax? ____________________

d. In each of these cases, how could you get them to return to their solid state?

________________________________________________________________

These three changes are examples of phase change because they have simply changed from one state of matter (phase) to another.

**Complete these activities.**

**2.25** Let's look at another example of a phase change. Get a candle and a short piece of solder. Light the candle and heat the solder in the flame.

a. What happens? ____________________________________________

b. How can you tell this change is a phase change? ____________________

________________________________________________________________

Physical properties are determined by the compound and are unique to that specific compound. No two compounds have the same identical properties. As a result, unknown substances can be identified by comparing their physical properties to those of standard known substances. The investigation of melting characteristics of PDCB is one example.

**Chemical.** When elements react to form compounds, not only do their physical properties change, but also their chemical properties change. We can study this change in properties by analyzing a classic example of the change. We shall use this example as a model to identify when **chemical changes** occur and how to distinguish between elements and compounds by changes in their properties.

When a magnet is brought near iron filings, the filings are attracted to the magnet. However, sulfur is not affected by the magnet. (See Box 1 in Figure 6.)

Carbon disulfide ($CS_2$) dissolves sulfur, but does not affect iron filings at all. (See Box 2 in Figure 6.)

When sulfur and iron filings are mixed together, they appear to lose their identity. A yellow-gray mixture results in which neither iron nor sulfur can be clearly identified. However, when a magnet is brought near the mixture, the iron filings are separated out and the sulfur is left behind. When the mixture is placed in a carbon disulfide, the sulfur dissolves leaving the iron behind in the bottom of the beaker. (See Box 3 in Figure 6.)

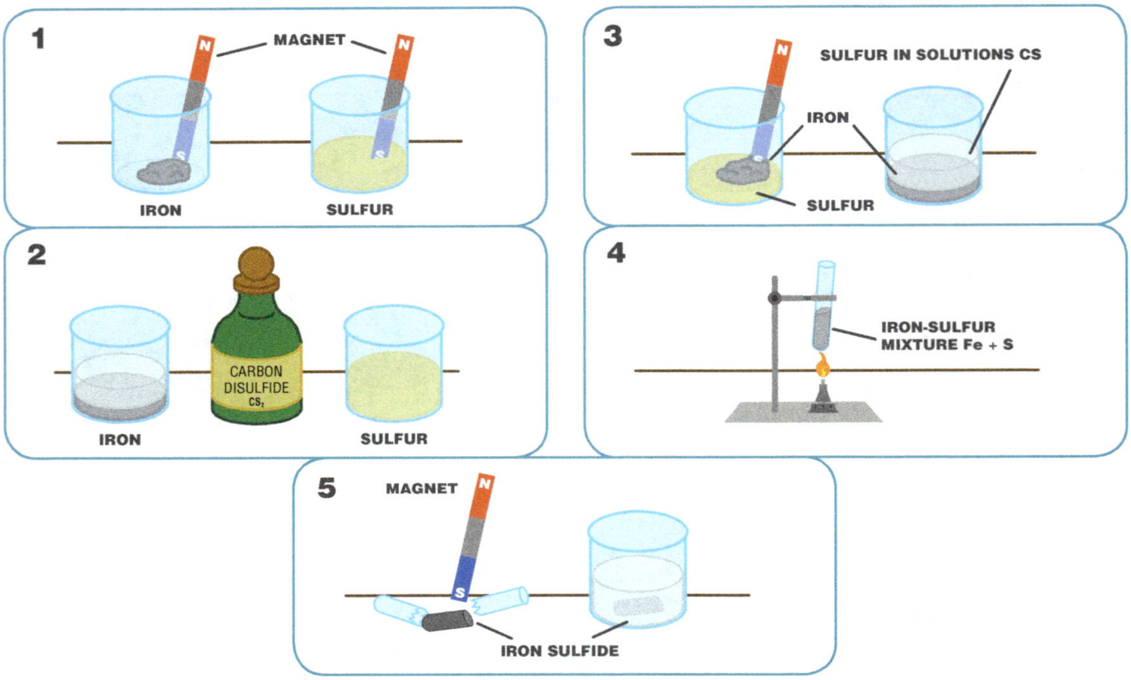


| Figure 6: Iron Sulfur Experiment

**Complete this activity.**

**2.26** Given a sulfur-iron mixture, propose a way to purify the sulfur and iron.

______________________________________________

______________________________________________

______________________________________________

______________________________________________

______________________________________________

______________________________________________

**Let's investigate.**

**These supplies are needed:**

- pure white sand
- pure salt
- a mixture of approximately ½ g sand and ½ g salt
- filter funnel
- filter paper or heavy paper hand towel

**Complete the investigation as you do each activity.**

**2.27** Describe pure sand and pure salt. Use a hand lens if available to get more detail. Draw pictures of each. Write a description.

a. Sand ________________________________

________________________________

________________________________

________________________________

________________________________

________________________________

b. Salt ________________________________

________________________________

________________________________

________________________________

________________________________

________________________________

**2.28** Investigate the dissolving properties of pure sand and pure salt. Which one will be caught in the filter paper if a water mixture of that compound is filtered?

________________________

Filter both separately if you are not sure.

**2.29** Devise a plan to get pure sand and pure salt in different piles from a 1 gram sample of the mixture.

a. My Plan: ________________________________________________

________________________________________________

________________________________________________

________________________________________________

**Separate a Mixture Experiment** (Continued on next page)

b. Results of my plan: ______________________________________________

______________________________________________

______________________________________________

______________________________________________

______________________________________________

______________________________________________

c. Grams dry sand ____________________

d. Grams dry salt ____________________

✓ **CHECK** ____________________ ____________________

Teacher Date

## Separate a Mixture Experiment

Let's return to our iron-sulfur experiment to further our ideas about the properties of elements and compounds in a chemical reaction. When the sulfur-iron mixture is heated in a test tube until the contents glow red, the appearance of the mixture changes. The heat is removed, and the mixture is allowed to cool. When the test tube is broken and the resulting substance is brought near a magnet, nothing happens. The heating has caused a change in the iron's magnetic properties and iron is not separated out by the magnet. When the heated iron-sulfur mixture is put into the carbon disulfide, no change results. No sulfur dissolves into the liquid carbon disulfide. We can conclude that something new has been formed. The old individual properties of both iron and sulfur have disappeared and new properties have appeared. A new substance has been formed with new properties. Chemically this new compound is the result of the uniting of sulfur and iron to form iron sulfide. A chemical reaction has occurred. (See Boxes 4 and 5 in Figure 6.)

Our spiritual life after accepting Jesus Christ as our Savior is much the same as a chemical reaction. The old individual characteristics have disappeared and a new person has been created. 2 Corinthians 5:17 states, "Therefore if a man be in Christ, he is a new creature: old things are passed away; behold, all things are become new."

| Chemical change?

Converting one set of chemicals into another is a basic operation of chemistry and life itself. Another example of a chemical change is called photosynthesis. Plants that have green pigment called chlorophyll are capable during daylight of taking carbon dioxide ($CO_2$), water ($H_2O$), and solar energy and making new plant materials and oxygen.

$CO_2 + H_2O +$ solar energy $\longrightarrow$ sugar, starch, cellulose $+ O_2$

Name two or three other examples of chemical change that you have experienced today.

# SCIENCE 1102

LIFEPAC TEST

**Name** ____________________

**Date** ____________________

**Score** ____________________

**78**

**98**

# SCIENCE 1102: LIFEPAC TEST

**Answer** *true* **or** *false* (each answer, 1 point).

**1.** __________ Energy is required to change a solid to a liquid.

**2.** __________ When a solid melts, heat is released.

**3.** __________ A chemical change occurs when water melts.

**4.** __________ Melting point is a physical characteristic that identifies an element.

**5.** __________ Rome was the seat of all alchemy.

**6.** __________ The chemical CuS is a mixture.

**7.** __________ When the reactants differ from the products, a chemical change has taken place.

**8.** __________ An element is two or more kinds of atoms fastened together.

**9.** __________ *Organic* means *from life*.

**10.** __________ $CaCO_3$ is an inorganic compound.



**Answer these questions** (each lettered answer, 2 points).

**11.** Indicate whether each of the following items is a chemical change or a phase change.

a. __________ Water boiling
b. __________ A bomb exploding
c. __________ Iron rusting
d. __________ Apple slice turning brown
e. __________ Burning paper
f. __________ Rotting trees

**12.** Indicate whether each of the following items is an inorganic or organic compound.

a. __________ $CaCO_3$
b. __________ $H_2O$
c. __________ $CH_4$
d. __________ $C_6H_{12}O_6$
e. __________ NaCl
f. __________ $C_2H_6O$

**13.** Indicate whether each of the following items is a pure substance or a mixture.

a. __________ soap
b. __________ pond water
c. __________ air
d. __________ fresh paradichlorobenzene
e. __________ oxygen from a bottle
f. __________ 2% milk

**14.** Indicate whether each of the following items is an element or compound.

a. __________ Co
b. __________ CO
c. __________ $CO_2$
d. __________ $CuCl_2$
e. __________ Ba
f. __________ HCl

**Circle the best answer for each of the following items** (each answer, 2 points).

Questions 15–20 refer to the warming curve for ice being changed to vapor.

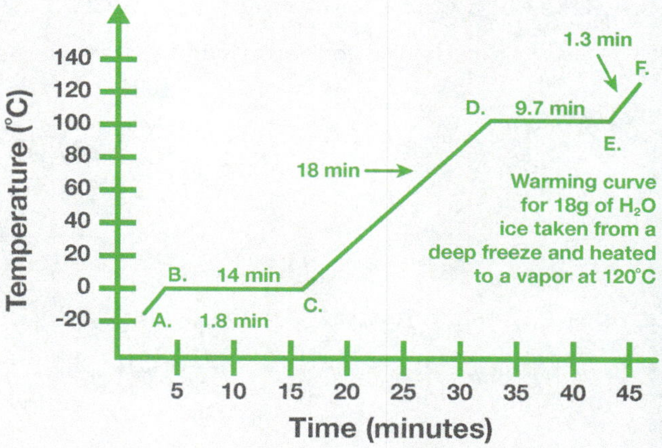


**15.** What point on the graph represents the lowest kinetic energy?

a. A  b. B
c. D  d. F

**16.** Where is the $H_2O$ being warmed as a liquid?

a. AB  b. BC
c. CD  d. DE

**17.** A phase change occurs.

a. AB  b. BC  c. CD  d. EF

**18.** What portion of the curve represents the greatest kinetic energy?

a. BC  b. CD  c. DE  d. EF

**19.** For the warming curve in the figure, the best explanation for the flat portion of the curve from B to C is __________ .

a. A liquid evaporating requires the energy which would normally be used to continue warming the water.

b. The heat from the flame that is warming the substance takes time to get to where the thermometer is, thus interrupting the heating.

c. Energy which could continue raising the temperature of the water is actually being lost to the surroundings.

d. Energy which could raise the temperature of the water is actually being transferred to potential energy to force the $H_2O$ particles apart.

**20.** Equal weight of ice and water are stored in a refrigerator that maintains them at exactly 0°C (32°F). You might expect __________ .

a. The ice to melt by using the extra kinetic energy available in the liquid.

b. The liquid to freeze by gaining potential energy from the ice.

c. The ice cube would melt and freeze at the same time and remain the same.

d. The ice cube would alternately freeze completely and then melt completely.

**Complete each of the following statements** (each lettered answer, 3 points).

**21.** All homogeneous matter can be classified as a(n) a. _______________ substance or a(n) b. _______________ .

**22.** Elements can be classified as a. _________________ or b. _________________ .

**23.** $H_2O$ contains the elements a. _______________ and b. _______________ (give names).

**Complete these activities** (each answer, 5 points).

**24.** Describe in detail what is meant by alchemy, what its purposes were, and what it contributed to today's world.

______________________________________________________________________

______________________________________________________________________

______________________________________________________________________

______________________________________________________________________

______________________________________________________________________

______________________________________________________________________

______________________________________________________________________

______________________________________________________________________

______________________________________________________________________

______________________________________________________________________

**25.** Define and contrast homogeneous and heterogeneous mixtures and give at least two examples of each.

______________________________________________________________________

______________________________________________________________________

______________________________________________________________________

______________________________________________________________________

______________________________________________________________________

______________________________________________________________________

______________________________________________________________________

______________________________________________________________________

______________________________________________________________________

______________________________________________________________________

______________________________________________________________________

______________________________________________________________________

Figure 7 illustrates another chemical reaction. When sulfuric acid is added to sugar, a chemical reaction occurs. The sugar turns black, a vapor is given off and the reaction gets very hot. When the black substance is analyzed, it is found to be the element carbon. The vapor is mostly water. The result is that the compound sugar can be broken down into carbon and water, quite different substances from the starting compound of sugar.

| Figure 7: Sugar-sulfuric acid reaction

We have already discussed that together the elements of oxygen and hydrogen form the compound water. Water can, in turn, be decomposed to form the elements of hydrogen and oxygen. This decomposition is done through a process called electrolysis, which is the process of passing an electrical current through water. Hydrogen is collected in one tube and oxygen in the other.

This water ⟷ hydrogen/oxygen reaction also illustrates another very important chemical concept. Water is always made from two parts hydrogen and one part oxygen by volume. Whatever amounts of hydrogen or oxygen are available, the reacting ratio is always 2:1. When water is decomposed, wherever the water comes from in the universe, it forms two volumes of hydrogen to one volume of oxygen. (See Figure 8.)

This reacting ratio is called the *Law of Constant Composition* or the *Law of Definite Proportions*. Clearly this law is a natural law designed by the Creator to give order and regularity to His creation, not something that developed from a chance occurrence in a random, chaotic universe. One might call this law God's Law of Regularity.

From our investigation so far, we may generalize and say that a chemical reaction changes the basic chemical nature of the material. In other words, the products (substances produced) are different in their makeup than the reactants (substances started with).

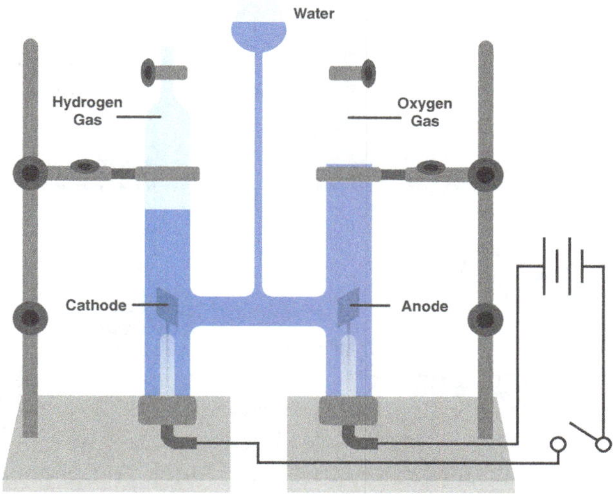


| Figure 8: Electrolysis of water

**Tell whether each of the following is an example of a phase change, physical change, chemical change or a combination. If a combination, tell which ones.**

**2.30** melting ice to water ____________________

**2.31** souring of milk ____________________

**2.32** digesting food ____________________

**2.33** evaporating water ____________________

**2.34** grinding glass ____________________

**2.35** rusting of iron ____________________

**2.36** burning a candle ____________________

**2.37** sanding wood ______________________

**2.38** frying eggs ______________________

**2.39** baking a cake ______________________

**CHECK** ______________________ ______________________
Teacher Date

**Choose one of the following optional activities. On separate paper, write a report and attach the report to this LIFEPAC.**

a. Prepare a detailed report on alternative energy sources for the future. Include such things as availability, length of use for that resource, cost, and environmental impact.

b. Determine the density of three pure solids on $g/cm^3$. Attach the report of procedure, results, and conclusions. Secure an unknown and determine which of the three solids it is. Discuss the question, "What factors cause differences in density?" Be specific.

c. Prepare a report on solid waste disposal, including such topics as garbage disposal, recycling, sources of energy, costs, and pollution problems. Discuss local problems and solutions to solid waste disposal.

## CLASSIFICATION

For purposes of study, compounds are divided into two categories based on association in nature. Those compounds found in organisms or derived from living sources are called organic compounds. All other compounds—specifically those not produced in life functions—are called inorganic compounds. The classification is not foolproof, and some compounds may be placed in both categories.

**Inorganic.** Inorganic compounds are those generally coming from the nonliving part of the earth. Compared to organic compounds, there are few kinds of inorganic compounds; but the total mass of inorganic matter in the universe far exceeds the mass of organic matter. Common examples of inorganic compounds are sand (silicon dioxide, $SiO_2$), table salt (sodium chloride, NaCl), limestone (calcium carbonate, $CaCO_3$), and water (dihydrogen oxide, $H_2O$). All the rocks and minerals of the lithosphere are inorganic compounds. Thousands of different inorganic compounds exist in our universe.

| Salt is an inorganic compound.

**Organic.** Organic compounds are derived from things that are or have been alive. They include all compounds which contain carbon, except the gases carbon dioxide ($CO_2$) and carbon monoxide (CO), and the compounds that include the carbonate group ($-CO_3$). All organic compounds contain carbon (C), nearly always hydrogen (H), and often oxygen (O) and nitrogen (N). Sometimes the elements of sulfur (S), phosphorous (P), magnesium (Mg), potassium (K), sodium (Na), and a few others are present. The biosphere is made up of organic compounds and includes all living creatures, their waste and decomposition products, and such fossilized materials as coal, oil shale, natural gas, and petroleum. Nearly all synthetic, plastic materials are made from petroleum or coal materials and are therefore organic compounds. The number of different organic compounds is in the hundreds of thousands and continues to grow steadily as new synthetics and plastics are made by scientists and researchers.

| Many products are made from refined oil (petroleum).

**Label each of the following compounds as inorganic** (*I*) **or organic** (*O*).

**2.40** ________ glucose ($C_6H_{12}O_6$)

**2.41** ________ salt (NaCl)

**2.42** ________ sodium carbonate ($Na_2CO_3$)

**2.43** ________ methane ($CH_4$)

**2.44** ________ DDT ($C_{14}H_9Cl_5$)

**2.45** ________ iron oxide ($Fe_2O_3$)

**2.46** ________ gasoline ($C_8H_{18}$)

**2.47** ________ vitamin $D_2$ ($C_{27}H_{40}O$)

**2.48** ________ mercury sulfate (HgS)

**2.49** ________ butane ($C_4H_{10}$)

**2.50** ________ sand ($SiO_2$)

**2.51** ________ water ($H_2O$)

**2.52** ________ styrene ($C_8H_8$)

**2.53** ________ adenine ($C_5H_4N_5$)

**2.54** ________ sulfuric acid ($H_2SO_4$)

**Review the material in this section in preparation for the Self Test.** The Self Test will check your mastery of this particular section as well as your knowledge of the previous section.

# SELF TEST 2

**Match these items** (each answer, 2 points).

| | | | |
|---|---|---|---|
| **2.01** | __________ | Pb | a. breaking glass |
| **2.02** | __________ | H | b. chlorine |
| **2.03** | __________ | physical change | c. lead |
| **2.04** | __________ | Cl | d. zerconium |
| **2.05** | __________ | phase change | e. $C_6H_{12}O_6$ |
| **2.06** | __________ | N | f. hydrogen |
| **2.07** | __________ | Cu | g. nitrogen |
| **2.08** | __________ | organic compound | h. $Na_3PO_4$ |
| **2.09** | __________ | Zn | i. melting ice |
| **2.010** | __________ | inorganic compound | j. digestion |
| | | | k. copper |
| | | | l. zinc |

**Complete these statements** (each answer, 3 points).

**2.011** The two classes of pure substances are a. ______________ and b. ______________ .

**2.012** In a chemical reaction, the starting substances are called a. ______________ and the substances resulting from that change are called b. ______________ .

**2.013** Homogeneous matter may be classified as a. ______________ or b. ______________ .

**2.014** Alchemy got its real beginning in ______________ .

**2 015** God is a God of order and harmony. A natural law that illustrates this statement in his creation is the ______________ .

**2.016** A spiritual application showing the loss of an individual's identity much as an element loses its identity in a compound is found in what Bible verse? ______________ .

**2.017** The energy of motion is called ______________

**Circle the letter of the best answer for each item** (each answer, 2 points).

The following graph of data was obtained from heating a sample of pure substance. The experiment is similar to the one you did in heating paradichlorobenzene. The steady heat source supplied 500 calories of heat per minute to the 60 gram sample. Use this information and the graph to answer questions 2.018 to 2.021.

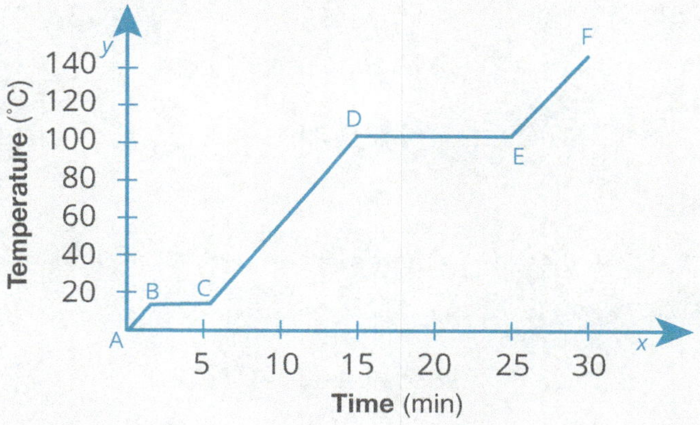


**2.018** The portion of the graph between B and C represents the time the substance is ______ .

a. being warmed as a solid
b. being warmed as a liquid
c. being warmed as a gas
d. changing from solid to liquid at its melting temperature
e. changing from liquid to gas at its boiling temperature

**2.019** The portion of the graph between C and D represents the time the substance is ______ .

a. being warmed as a solid
b. being warmed as a liquid
c. being warmed as a gas
d. changing from solid to liquid at its melting temperature
e. changing from liquid to gas at its boiling temperature

**2.020** The portion of the graph representing the greatest motion of molecules is ______ .

a. A–B
b. B–C
c. C–D
d. D–E
e. E–F

**2.021** The melting point of the substance is closest to ______ .

a. 0°C
b. 5°C
c. 15°C
d. 65°C
e. 115°C

**Complete these activities** (each answer, 2 points).

**2.022** Indicate whether each of the following items is a *chemical* (*C*) or *physical* (*P*) change.

a. _____ exploding firecracker
b. _____ mowing grass
c. _____ breaking a window
d. _____ baking a cake
e. _____ souring of milk
f. _____ taking Rolaids® for acid stomach

**2.023** Classify each item as *organic* (*O*) or *inorganic* (*I*).

a. _____ $NaCl$
b. _____ $C_2H_6$
c. _____ formaldehyde ($CH_2O$)
d. _____ phenol ($C_6H_6O$)
e. _____ $Ca_3(PO_4)_2$
f. _____ $MgCO_3$

**2.024** Classify each of the following items as a *compound (C)* or an *element (E)*.

a. _____ $HCl$
b. _____ $O_2$
c. _____ $S_8$
d. _____ Co
e. _____ $H_2SO_4$
f. _____ $N_2$

**Define each of the following words** (each answer, 3 points).

**2.025** reactant ________________________________________________

________________________________________________

________________________________________________

**2.026** compound ________________________________________________

________________________________________________

________________________________________________

80/100 **SCORE** __________ **CHECK** __________ Teacher __________ Date

# 3. MIXTURES

Much of the chemistry we experience in our everyday life involves a mixture of substances. The following diagram should help to review and to describe mixtures.

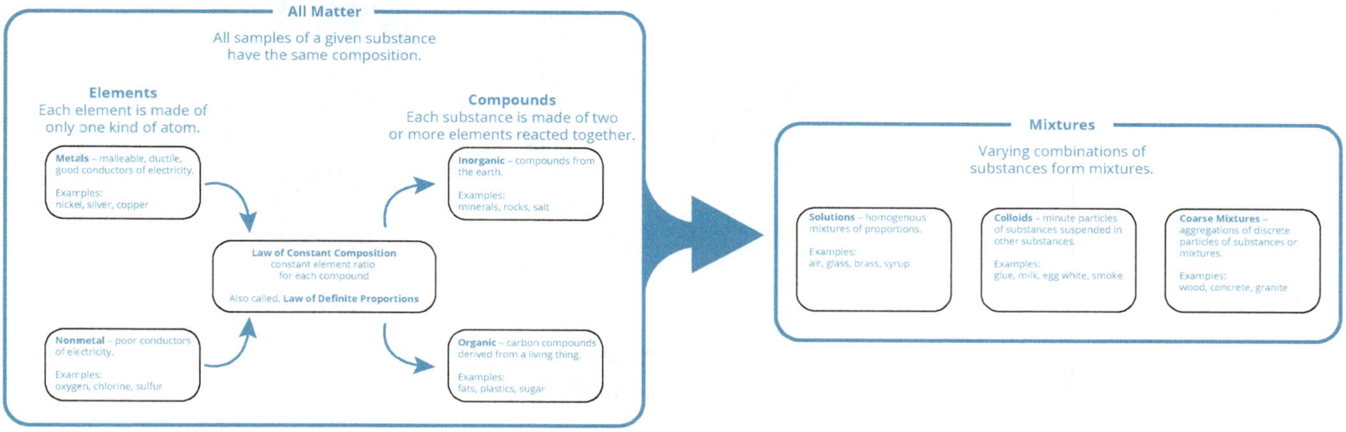


| Figure 9

## Section Objectives

**Review these objectives.** When you have completed this section, you should be able to:

7. Describe and give examples of various types of mixtures.
   - 7.1 Describe and give examples of heterogeneous mixtures.
   - 7.2 Describe and give examples of homogeneous mixtures.

## Vocabulary

**Study these words to enhance your learning success in this section.**

**colloid** **heterogeneous** **homogeneous** **mixture** **solution** **suspension**

Life itself functions on the basis of **mixtures**. Our blood is a liquid mixture. Many of the foods we eat are mixtures. Peanut butter, milk, mayonnaise, gelatin, marshmallows, and whipped cream are just a few examples. Rubber, paints, cement, ceramics, and detergents are other mixtures that we experience daily. This section of the LIFEPAC will help us to understand some of the properties of mixtures.

# HETEROGENEOUS

Mixtures can be classified as heterogeneous or homogeneous. The primary basis for placement in one class or the other is particle size. Particle size can be determined by two basic properties:

1. Tyndall effect and
2. Brownian movement.

**Suspensions.** Large particles suspended in a mixture will scatter a light beam passed through the solution producing a phenomenon called the Tyndall effect. Since the particles are so large, the light is interrupted and bounced around when it hits the large, suspended particles. John Tyndall (1820-1893) was the first to notice and to study the phenomenon in **solutions**. Heterogeneous mixtures show the Tyndall effect. (See Figure 10.) A flashlight beam in a smoky room is an example of the Tyndall effect. Car or street light beams on a foggy night is another.

| Figure 10: Tyndall effect

The second characteristic of heterogeneous mixtures is to show Brownian movement. Robert Brown (1773-1858) discovered a unique characteristic of large particles suspended in a mixture. He was studying under a microscope pollen suspended in water. To his amazement, the pollen grains were moving around in a random zigzag manner. What he was witnessing was the effect of collisions between the water molecules and the pollen grains. The random collision between the water molecules and pollen grains caused the pollen to be pushed this way and that, much as a person might be pushed around in a big crowd excited about a football game. Large particles in a heterogeneous mixture display the Brownian movement. (See Figure 11.) When the particles are very large and block the light completely and settle out of the dissolved state, if left to set, the system is called a **suspension**. (See Figure 11.)

**Colloids.** The amazingly large number of chemical reactions that are involved in food preparation qualifies "kitchen chemistry" as a logical extension of the chemical theory we will study this year. Some of the examples of chemistry that have immediate relevance to the kitchen include these: energy changes and oxidation in baking and fermentation, acid-base reactions in all cooking and bodily processes, precipitation in cheese and cottage cheese making, structure of foods including vitamins and minerals, chemical properties of foods which affect their preparation and flavor, chemical preservation of foods, and the list could continue.

In an attempt to simplify, we will limit the present investigation to the study of the particle size associated with **colloids**. Feel free to continue an investigation on your own if you think of ideas to test. We shall investigate by both the cookbook and the experimental-type inquiries, although some of your cooking might be quite experimental. Some of the experiments will be open-ended, and others will ask you to draw definite conclusions.

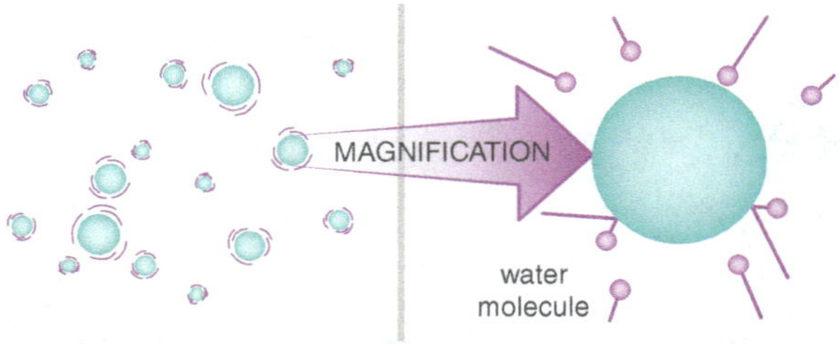


| Figure 11: Brownian Movement

When you are ready for the kitchen, don an apron and proceed to the pans (beakers) and food supply in the lab.

Particle size is important in cooking, and it may range from such atomic size as particles in a solution of salt in water to such visible particles as a heterogeneous mixture of clay suspended in water. The salt solution is an example of a homogeneous mixture which will not settle upon standing; but the mud suspension will, of course, settle. Kitchen chemistry often deals with "in-between" particles of colloidal size.

**Let's investigate.**

**These supplies are needed:**

- vegetable oil
- mustard
- vinegar
- homogenized milk
- water
- sour milk
- egg
- salt
- blender
- canned whipped topping (optional)
- large beaker or bowl
- canned shaving foam (optional)
- baby food jars
- sugar

**Complete these investigation activities.**

**3.1** To get an idea of colloidal size, try mixing some vegetable oil and vinegar in a test tube. Diagram and label the results at the right.

**3.2** Repeat the process of 3.1 with oil and water. What are the results?

______________________________

______________________________

______________________________

______________________________

______________________________

______________________________

______________________________

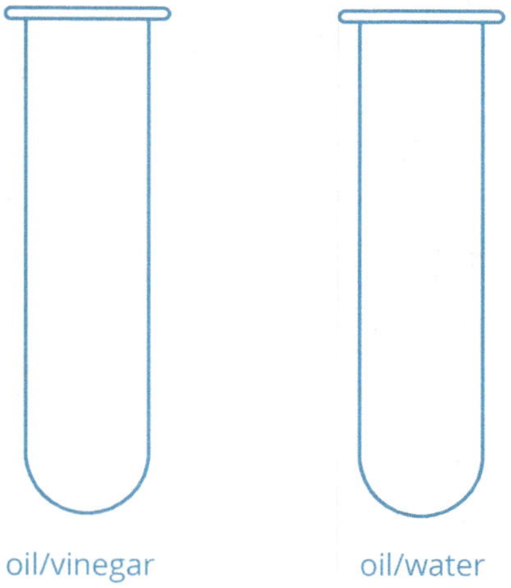


**Answer these questions.**

**3.3** a. Would you say that both mixtures in 3.1 and 3.2 are homogeneous or heterogeneous?

__________

b. Is a complete mixture possible? __________

c. Why or why not? (Use reference if needed.) ______________________________

______________________________

______________________________

______________________________

______________________________

**Colloids Experiment** (Continued on next page)

**Complete this investigation, then answer the question.**

**3.4** If the oil particles could be made to stay small, then the two substances might show different mixture properties. Obtain a blender, vinegar, and oil. Break an egg into the blender and mix it with a little salt, sugar, and mustard for flavoring. Slowly add 15 ml of vinegar while stirring the mixture vigorously with the blender. Gradually add 15 ml of vegetable oil with vigorous blending. Add seven successive 15 ml portions of vegetable oil, with continuous and vigorous blending.

*Observations*: ______________________________

______________________________

______________________________

**3.5** Would you judge your mixture from 3.4 to have become more homogeneous or more heterogeneous? ____________ You should have produced a suspension. If your suspension starts to separate (break), then you are finished, unless you can save it by slowly adding 15 ml of cold water or egg white with slow mixing. The final mixture—called mayonnaise—is an emulsion or suspension of vegetable oil in vinegar with egg as the stabilizing agent. Other examples of this type of suspension are French dressing, milk, and cake batter.

**Answer these questions.**

**3.6** Why is the suspension produced in 3.4 and 3.5 different from the attempted mixture in the previous activities, 3.1 and 3.2?

______________________________

______________________________

______________________________

______________________________

**3.7** a. Is mayonnaise a solution? ____________

b. Explain briefly. ______________________________

______________________________

______________________________

**Complete these investigation activities.**

**3.8** Now try emulsifying oil in water by using a stabilizing agent of your choice. Dirty dishes or clothes are cleaned in this way. Explain your procedure and the results.

a. Procedure: ______________________________

______________________________

______________________________

**Colloids Experiment** (Continued on next page)

b. Results: ____________________________________________

____________________________________________

____________________________________________

____________________________________________

**3.9** The preceding examples of emulsions involve breakdown of particle size. Another example is the homogenization of milk, a process which involves suspending

a. ______________ particles in a medium of b. ______________ .

**Complete these investigation steps and answer the questions.**

**3.10** The opposite process can be observed.

a. Place 2 ml of homogenized milk in a test tube. Slowly add, with shaking, about 10 drops of lemon juice or acetic acid (vinegar).

What appears to happen? ____________________________________________

____________________________________________

____________________________________________

____________________________________________

b. Centrifuge the products. If the milk was a suspension to start with, what did the addition of acid do to the size of the suspended particles? (Note: If you don't have a centrifuge, see your teacher.)

____________________________________________

____________________________________________

____________________________________________

c. Observe and centrifuge some sour milk. How does it compare with the product in the previous question?

____________________________________________

____________________________________________

____________________________________________

**Colloids Experiment** (Continued on next page)

**Answer these questions.**

**3.11** a. Would milk fresh out of a cow exhibit the same physical properties as the homogenized milk in the carton? ________________

b. Why or why not? ______________________________________________

______________________________________________

______________________________________________

**3.12** What did you learn about the three different particle sizes and how they can be changed? Summarize in a sentence or two.

______________________________________________

______________________________________________

______________________________________________

______________________________________________

______________________________________________

✓ **CHECK** ________________ ________________
Teacher Date

## Colloids Experiment

The following diagram of decreasing particle size will help clarify the three types of particles in relation to their physical properties.

| Mixture Type | Particle Size | Examples | Light Beam |
|---|---|---|---|
| **Suspensions** (heterogeneous) (temporary) | $10^{-6}$ cm | Coagulated milk, clay in $H_2O$ | Interrupted |
| **Colloids** (heterogeneous) (semipermanent) | $10^{-6}$ – $10^{-8}$ cm | Milk, batter, egg white, foam, mayonnaise, gels | Visible within |
| **Solutions** (homogeneous) (permanent) | $10^{-8}$ cm<br>Decreasing size | Salt in $H_2O$, plasma, saliva, ammonia in $H_2O$ | Unaffected |

**You may do this optional activity.**

OPTIONAL: Spray some canned whipped topping or canned shaving foam on a glass plate. What kind of a suspension do you have; i.e., what is suspended in what? Explain. Attach a report on the optional activity.

In its natural state, rubber is a colloid, and its preparation for use depends upon principles of colloidal chemistry. The manufacturing of leather from hides is a colloidal process. Glues and adhesives are colloids. The preparation of soils for agricultural purposes, the manufacture of photographic plates and films, the silvering of glass for mirrors, certain methods of dyeing, the flotation process used in metallurgy, and the manufacturing of explosives—all of these processes involve a knowledge of the properties of colloids.

| Type | Examples |
|---|---|
| Solid-in-liquid | Gelatin |
| Liquid-in-liquid (emulsion) | Homogenized milk |
| Gas-in-liquid (foam) | Whipped cream |
| Solid-in-solid | Black diamond |
| Gas-in-solid | Marshmallow |
| Solid-in-gas | Smoke |
| Liquid-in-gas (aerosol) | Fog |

| Figure 12: Examples of Colloidal Systems

**Define these terms.**

**3.13** Both suspensions and colloids are heterogeneous mixtures. Define and characterize each, listing similarities and differences. Give several examples of each.

a. suspension ________________________________________________

b. colloid ________________________________________________

# HOMOGENEOUS

We have studied liquid heterogeneous mixtures (solutions) and found that particle size determines the characteristics with regard to the Tyndall effect and Brownian movement. When particle size is less than $10^{-8}$ cm neither test is positive. Light passes directly through the liquid mixture uninterrupted and no Brownian movement is evidenced even under high magnification. This type of mixture is called a true solution. Other types of homogeneous mixtures exist between other states of matter. For example, a homogeneous solid mixture is peanut butter.

**Liquid Solutions.** True solutions are difficult to separate into their individual components by ordinary means. Salt water or sugar water can be separated by evaporation of the water, leaving the salt or sugar behind with the water being condensed in a pure state.

Sometimes a mixture of pure substances is dissolved in a liquid (usually water) and a true solution results. This type of situation presents a difficult separation task. Blood and plasma analyses are examples of the need to separate individually pure substances into their true water solutions which are made from a mixture of substances. Ordinary evaporation techniques will only result in the separation of the water and will not produce the individual substances in their pure state. One technique used to separate homogeneous liquid mixtures is the technique of *chromatography*.

**Let's investigate a simple chromatography system.**

**These supplies are needed:**

- clean, washed, white sand (dry)
- graduated cylinder (100 ml)
- at least 15 small baby food jars or clear plastic 8 oz cups
- clear plastic or glass tube (at least 30 cm long)
- grape powdered drink mix
- 1 hole stopper to fit tube
- cloth surgical gauze
- short pipe or glass tube (fits hole in stopper)

**Set up the following system.**

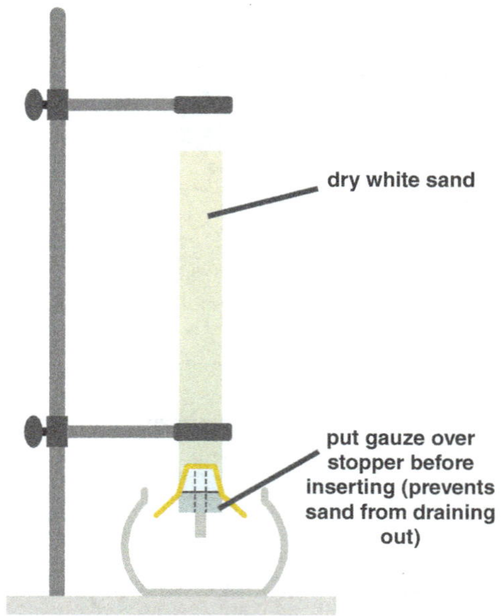


**Chromatography Experiment** (Continued on next page)

**Follow these directions. Put a check mark in the box when each step is completed.**

- ☐ 1. Place a baby food jar under the drip tube and pour 100 ml of water into the column to wet the sand. If necessary, add more water until the whole column is thoroughly wet.
- ☐ 2. When dripping is completed, discard the water that drips out of the filtering system.
- ☐ 3. Number the baby food jars from 1 to 15.
- ☐ 4. Place a pinch of grape powdered drink mix on the top of the sand column.
- ☐ 5. Add 100 ml portions of water, catching each in separate jars and allowing the dripping to stop between each successive new water addition.
- ☐ 6. Carefully observe the movement of the drink mix down through the column noting any color changes or separations.
- ☐ 7. Using a grease pencil or marking pen, place a line on the clear column, marking with each successive 100 ml addition the furthest point of downward movement of the drink mix.
- ☐ 8. When the dripping has stopped after each successive 100 ml additions, measure the distance from the top of the sand to that mark.
- ☐ 9. Continue this process until the drip water is again clear. Use additional jars if necessary and extend the Data Table accordingly.

**Record the data from your investigation and all observations in the following Data Table 3.**

**3.14** Fill in the first three columns. You will complete the last column after you are finished with all trials and have read 3.15.

**Data Table 3**

| Trial | Distance Moved (cm) | Observations | Color of Dripped Water | Relative Color Intensity |
|---|---|---|---|---|
| Jar 1 | | | | |
| Jar 2 | | | | |
| Jar 3 | | | | |
| Jar 4 | | | | |
| Jar 5 | | | | |
| Jar 6 | | | | |
| Jar 7 | | | | |
| Jar 8 | | | | |
| Jar 9 | | | | |
| Jar 10 | | | | |
| Jar 11 | | | | |
| Jar 12 | | | | |
| Jar 13 | | | | |
| Jar 14 | | | | |
| Jar 15 | | | | |

✓ **CHECK** ____________ ____________
Teacher Date

**Chromatography Experiment** (Continued on next page)

**Complete these activities about the preceding investigation.**

**3.15** Line up all of the jars in numerical order. Assign the darkest one a value of 10 and record that value for that jar in the last column of Data Table 3. Rate each of the other jars from 1 to 10 based on comparative color intensities with 1 being colorless.

a. Record each in Data Table 3.

b. On the following graph, plot your intensity ratings and jar numbers.

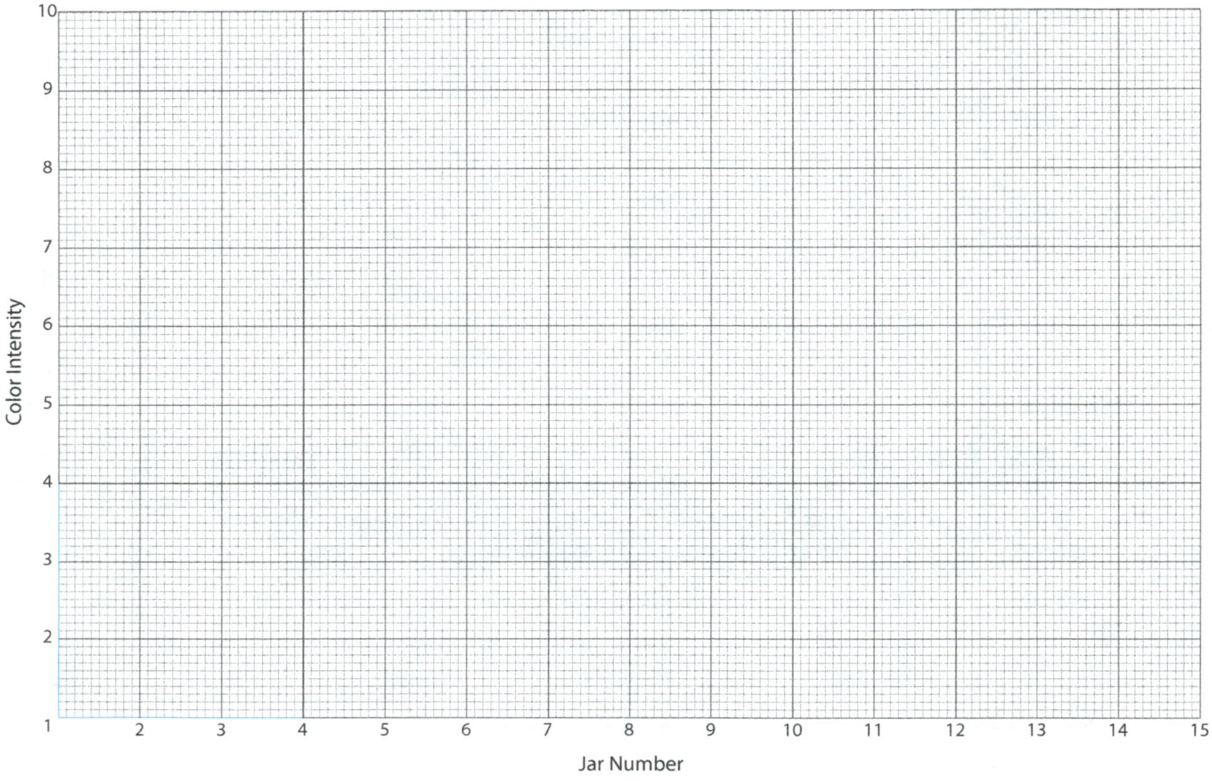


**Answer these questions.**

**3.16** a. Was each of the contents of the jars the same color? ________________

b. Did each color move through the sand at the same rate? ________________

c. What do you conclude? ________________________________________________

________________________________________________

________________________________________________

**Chromatography Experiment**

(Continued on next page)

**3.17** Pesticides, herbicides, and fertilizers decompose to form a variety of soil pollutants. These pollutants build up, leach out, and then move through the soil; ending up in lakes, streams, oceans, and underground water supplies. On the basis of this investigation, tell the dangers and results of excess use of these chemicals, and what problems result from the time lag between the continued usage and amount present in the water supplies.

______________________________________________

______________________________________________

______________________________________________

______________________________________________

______________________________________________

______________________________________________

______________________________________________

______________________________________________

______________________________________________

______________________________________________

**You may complete this optional activity.**

OPTIONAL: The field of chromatography is both interesting and challenging. Attach a report on how it works, its applications, and what its future potential may be as a technical aid to chemists and other scientists.

## Chromatography Experiment

**Gas Solutions.** One of the most obvious homogeneous mixtures is air. It is a mixture of gases that is uniformly mixed in such a way that anywhere you go on earth, the same proportion of each gas will be present. The only variable will be due to man's influence and most frequently is the result of man's pollution. Most solid pollutants settle out, allowing the air to become pure and clean again. This settling occurs naturally without any effort or act of man. Our Creator has provided by the law of gravity, a way for the air to become clean.

The natural composition of air is illustrated in the chart at the right.

| Substance | % Abundance |
|---|---|
| Nitrogen ($N_2$) | 78.08000 |
| Oxygen ($O_2$) | 20.95000 |
| Argon (Ar) | 0.93400 |
| Carbon Dioxide ($CO_2$) | 0.03140 |
| Neon (Ne) | 0.00182 |
| Others | 0.10278 |

**Complete these activities.**

**3.18** What is the most abundant gas in our atmosphere? ____________________

**3.19** On the basis of information presented in this section about mixtures and air, describe the characteristics of gaseous mixtures. Be specific and use as much data as possible.

____________________
____________________
____________________
____________________
____________________
____________________
____________________
____________________
____________________

**CHECK** ____________________ ____________________
Teacher Date

**Before you take this last Self Test, you may want to do one or more of these self checks.**

1. ______ Read the objectives. Determine if you can do them.
2. ______ Restudy the material related to any objectives that you cannot do.
3. ______ Use the **SQ3R** study procedure to review the material:
   a. **S**can the sections.
   b. **Q**uestion yourself again (review the questions you wrote initially).
   c. **R**ead to answer your questions.
   d. **R**ecite the answers to yourself.
   e. **R**eview areas you did not understand.
4. ______ Review all vocabulary, activities, and Self Tests, writing a correct answer for every wrong answer.

# SELF TEST 3

**Answer** *true* **or** *false* (each answer, 1 point).

3.01 ____________ Energy is not required to change a solid to a liquid.

3.02 ____________ Heat is released when a substance changes from a liquid to a solid.

3.03 ____________ A change that occurs when a solid melts or a liquid freezes is an example of a phase change.

3.04 ____________ The temperature at which a solid melts is not characteristic of a compound nor an element.

3.05 ____________ When products are different from reactants, a physical change has occurred.

3.06 ____________ The melting behavior is one of the most commonly used methods of characterizing a substance.

3.07 ____________ When heat is absorbed by water at room temperature, the temperature of the water falls.

3.08 ____________ A solid is in a higher energy state than the same chemical as a liquid.

3.09 ____________ Jerusalem was the site of the first study of alchemy.

3.010 ____________ The chemical $CuSO_4$ is an element.

**Match these items** (each answer, 2 points).

3.011 ____________ Cu

3.012 ____________ atom

3.013 ____________ organic compound

3.014 ____________ H

3.015 ____________ S

3.016 ____________ element

3.017 ____________ inorganic compound

3.018 ____________ mixture

3.019 ____________ compound

3.020 ____________ Fe

a. hydrogen
b. one substance of two or more elements
c. iron
d. two or more pure substances together
e. selenium
f. calcium
g. smallest unit retaining properties of element
h. NaCl
i. copper
j. $C_6H_6$
k. only one substance
l. sulfur

**Questions 3.021–3.027 use the following experiment and data graph.**

A student heated a test tube of solid material. The student recorded both the time and the temperature for the material. He then plotted the graph shown.

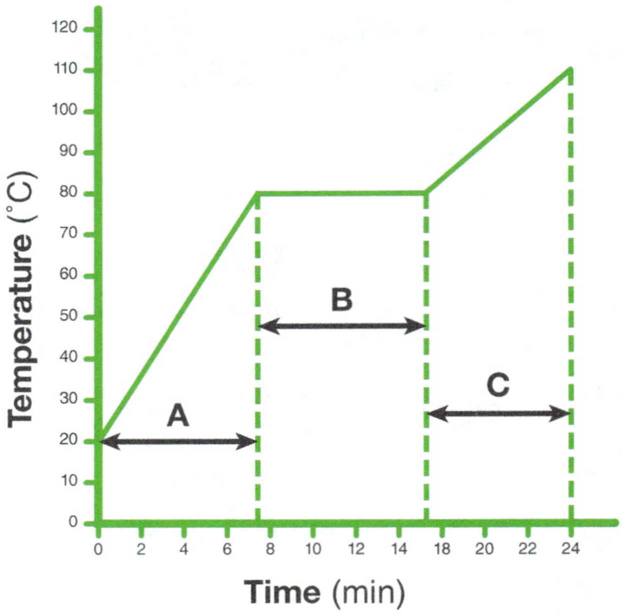


**Circle the best answer** (each answer, 2 points).

**3.021** During which time interval or intervals is there only liquid in the tube? __________

a. A only b. B only c. C only d. A and B only

**3.022** The melting point of the material is __________ .

a. 110°C b. 80°C c. 90°C
d. 0° C e. room temperature

**3.023** The freezing point of the material is __________ .

a. 0°C b. 110°C
c. 80°C d. room temperature

**3.024** During which time interval or intervals are both solid and liquid in the test tube? __________

a. C only b. A only c. A and B only d. B only

**Write** *decreasing*, *increasing*, **or** *constant* **in the blank** (each answer, 2 points).

**3.025** During the time interval 16–20 min., kinetic energy is ________________________________ .

**3.026** During the time interval 0–7 min., kinetic energy is ________________________________ .

**3.027** During the time interval 7.2–15.2 min., potential energy is ____________________________ .

Problems 3.028 and 3.029 use the following information:

Given: Closed container A with liquid water at 60°C
Closed container B with ice at 0°C
Closed container C with water vapor 100°C

**Choose the best answer** (each answer, 2 points).

**3.028** The molecules in container A as compared to the molecules in container B

a. move slower.
b. have more kinetic energy.
c. have less kinetic energy.
d. are separated by shorter distances.
e. a, b, c, and d.

**3.029** In which of the three containers do the molecules have the greatest kinetic energy?

a. C
b. A
c. B

**3.030** Which of the following items is a chemical change?

a. burning
b. melting
c. evaporation
d. freezing

**3.031** Which of the following sentences describes a physical change?

a. sugar dissolves in coffee
b. gasoline burns in an engine
c. iron rusts
d. a candle burns

**Complete each of the following items** (each answer, 2 points).

**3.032** Classify each item as *organic (O)* or *inorganic (I)*.

a. ______ table salt (NaCl)

b. ______ TNT ($C_7H_5N_3O_9$)

c. ______ glucose ($C_6H_{12}O_6$)

d. ______ 2, 4-D ($C_3H_6O_3Cl_2$)

e. ______ limestone ($CaCO_3$)

f. ______ water ($H_2O$)

**3.033** Classify each item as a compound (C) or element (E).

a. ______ Cu

b. ______ $CCl_4$

c. ______ Co

d. ______ HCl

e. ______ Hf

f. ______ HI

g. ______ $CH_4$

**Complete each of the following items** (each answer, 3 points).

**3.034** Two tests used to determine the sizes of particles in liquid mixtures are

a. ______________________ and b. ______________________ .

**3.035** The two main classes of mixtures are a. ________________ and b. ________________ .

**Complete each item using complete sentences** (each answer, 5 points).

**3.036** Describe in some detail what is meant by alchemy and what it has contributed to us today.

____________________________________________________________

____________________________________________________________

____________________________________________________________

____________________________________________________________

____________________________________________________________

____________________________________________________________

____________________________________________________________

**3.037** Define and contrast the two types of mixtures and give examples of each.

______________________________________________

80/100 **SCORE** ____________ **CHECK** ____________ ____________
Teacher Date

**Before taking the LIFEPAC Test, you may want to do one or more of these self checks.**

1. ________ Read the objectives. Determine if you can do them.
2. ________ Restudy the material related to any objectives that you cannot do.
3. ________ Use the **SQ3R** study procedure to review the material.
4. ________ Review activities, Self Tests, and LIFEPAC vocabulary words.
5. ________ Restudy areas of weakness indicated by the last Self Test.

# PERIODIC TABLE OF

Percent Ionic Character of a Single Chemical Bond

| Difference in electronegativity | 0.1 | 0.2 | 0.3 | 0.4 | 0.5 | 0.6 | 0.7 | 0.8 | 0.9 | 1.0 | 1.1 | 1. |
|---|---|---|---|---|---|---|---|---|---|---|---|---|
| Percent ionic character % | 0.5 | 1 | 2 | 4 | 6 | 9 | 12 | 15 | 19 | 22 | 26 | 3 |

| | Electron | Proton | Neutron |
|---|---|---|---|
| Symbol | e-,B- | +p | +n |
| Mass | 1 | 1836.12 | 1838.65 |
| Charge | -1 | +1 | 0 |
| Spin | 1/2 | 1/2 | 1/2 |

SYMBOL (2) — H<br>ATOMIC RADIUS, A (7) — —<br>FIRST IONIZATION ENERGY (kcal/g-mole) — 313<br>NAME — Hydrogen

H 1 – 1.009<br>— 1 -<br>313 2.1<br>$1s^{1}$ —<br>Hydrogen

# KEY

| IA | IIA | IIIB | IVB | VB | VIB | VIIB | VIII | VIII |
|---|---|---|---|---|---|---|---|---|
| **H** 1<br>1.00797<br>— 1<br>313 2.1<br>$1s^{1}$<br>Hydrogen | | | | | | | | |
| **Li** 3<br>6.939<br>1.55 1<br>124 1.0<br>$1s^{2}2s^{1}$<br>Lithium | **Be** 4<br>9.0122<br>1.12 2<br>215 1.5<br>$1s^{2}2s^{2}$<br>Beryllium | | | | | | | |
| **Na** 11<br>22.9898<br>1.90 1<br>119 0.9<br>$[Ne]3s^{1}$<br>Soduim | **Mg** 12<br>24.312<br>1.60 2<br>176 1.2<br>$[Ne]3s^{2}$<br>Magnesium | | | | | | | |
| **K** 19<br>39.102<br>2.35 1<br>100 0.8<br>$[Ar]4s^{1}$<br>Potassium | **Ca** 20<br>40.08<br>1.97 2<br>141 1.0<br>$[Ar]4s^{2}$<br>Calcium | **Sc** 21<br>44.956<br>1.62 3<br>151 1.3<br>$[Ar]3d^{1}4s^{2}$<br>Scandium | **Ti** 22<br>47.90<br>1.47 4,3<br>158 1.5<br>$[Ar]3d^{2}4s^{2}$<br>Titanium | **V** 23<br>50.942<br>1.34 5,4,3,2<br>156 1.6<br>$[Ar]3d^{3}4s^{2}$<br>Vanadium | **Cr** 24<br>51.996<br>1.30 6,3,2<br>156 1.6<br>$[Ar]3d^{5}4s^{1}$<br>Chromium | **Mn** 25<br>54.938<br>1.35 7,6,4,2,3<br>171 1.5<br>$[Ar]3d^{5}4s^{2}$<br>Manganese | **Fe** 26<br>55.847<br>1.26 2,3<br>182 1.8<br>$[Ar]3d^{6}4s^{2}$<br>Iron | **Co** 27<br>58.933<br>1.25 2,3<br>181 1.8<br>$[Ar]3d^{7}4s^{2}$<br>Cobalt |
| **Rb** 37<br>85.47<br>2.48 1<br>96 0.8<br>$[Kr]5s^{1}$<br>Rubidium | **Sr** 38<br>87.62<br>2.15 2<br>131 1.0<br>$[Kr]5s^{2}$<br>Strontium | **Y** 39<br>88.905<br>1.78 3<br>152 1.3<br>$[Kr]4d^{1}5s^{2}$<br>Yttrium | **Zr** 40<br>91.22<br>1.60 4<br>160 1.4<br>$[Kr]4d^{2}5s^{2}$<br>Zirconium | **Nb** 41<br>92.906<br>1.46 5,3<br>156 1.6<br>$[Kr]4d^{4}5s^{1}$<br>Nobium | **Mo** 42<br>95.94<br>1.39 6,5,4,3,2<br>166 1.8<br>$[Kr]4d^{5}5s^{1}$<br>Molybdenum | **Tc** 43<br>(98)<br>1.36 7<br>167 1.9<br>$[Kr]4d^{5}5s^{2}$<br>Technetium | **Ru** 44<br>101.07<br>1.34 2,3,4,6,8<br>173 2.2<br>$[Kr]4d^{7}5s^{1}$<br>Ruthenium | **Rh** 44<br>102.905<br>1.34 2,3,4<br>178 2.2<br>$[Kr]4d^{8}5s^{1}$<br>Rhodium |
| **Cs** 55<br>132.905<br>2.67 1<br>90 0.7<br>$[Xe]6s^{1}$<br>Cesium | **Ba** 56<br>137.34<br>2.22 2<br>120 0.9<br>$[Xe]6s^{2}$<br>Barium | | **Hf** 72<br>178.49<br>1.67 4<br>127 1.3<br>$[Xe]4f^{14}5d^{2}6s^{2}$<br>Hafnium | **Ta** 73<br>180.948<br>1.49 5<br>138 1.5<br>$[Xe]4f^{14}5d^{3}6s^{2}$<br>Tantalum | **W** 74<br>183.85<br>1.41 6,5,4,3,2<br>184 1.7<br>$[Xe]4f^{14}5d^{4}6s^{2}$<br>Wolfram | **Re** 75<br>186.2<br>1.37 7,6,4,2,-1<br>182 1.9<br>$[Xe]4f^{14}5d^{5}6s^{2}$<br>Rhenium | **Os** 76<br>190.2<br>1.35 2,3,4,6,8<br>201 2.2<br>$[Xe]4f^{14}5d^{6}6s^{2}$<br>Osmium | **Ir** 77<br>192.2<br>1.36 2,3,4,6<br>212 2.2<br>$[Xe]4f^{14}5d^{7}6s^{2}$<br>Iridium |
| **Fr** 87<br>(223)<br>— 1<br>— 0.7<br>$[Rn]7s^{1}$<br>Francium | **Ra** 88<br>(226)<br>— 2<br>— 0.9<br>$[Rn]7s^{2}$<br>Radium | | **Rf** 104<br>(261)<br>Rutherfordium | **Db** 105<br>(262)<br>Dubnium | **Sg** 106<br>(266)<br>Seaborgium | **Bh** 107<br>(262)<br>Bohrium | **Hs** 108<br>(263)<br>Hassium | **Mt** 109<br>(268)<br>Meitnerium |

| | | | | | | |
|---|---|---|---|---|---|---|
| **La** 57<br>138.91<br>1.87 3<br>129 1.1<br>$[Xe]5d^{1}6s^{2}$<br>Lathanum | **Ce** 58<br>140.12<br>1.81 3,4<br>159 1.1<br>$[Xe]4f^{2}5d^{0}6s^{2}$<br>Cerium | **Pr** 59<br>140.907<br>1.82 3,4<br>133 1.1<br>$[Xe]4f^{3}5d^{0}6s^{2}$<br>Praseodymium | **Nd** 60<br>144.24<br>1.82 3<br>145 1.2<br>$[Xe]4f^{4}5d^{0}6s^{2}$<br>Neodymium | **Pm** 61<br>(147)<br>— 3<br>133 —<br>$[Xe]4f^{5}5d^{0}6s^{2}$<br>Promethium | **Sm** 62<br>150.35<br>1.81 3,2<br>129 1.2<br>$[Xe]4f^{6}5d^{0}6s^{2}$<br>Samarium | **Eu** 63<br>151.96<br>1.99 3,2<br>131 —<br>$[Xe]4f^{7}5d^{0}6s^{2}$<br>Europium |
| **Ac** 89<br>(227)<br>1.88 3<br>— 1.1<br>$[Rn]6d^{1}7s^{2}$<br>Actinium | **Th** 90<br>232.038<br>1.80 4<br>— 1.3<br>$[Rn]5f^{0}6d^{2}7s^{2}$<br>Thorium | **Pa** 91<br>(231)<br>1.61 5,4<br>— 1.5<br>$[Rn]5f^{2}6d^{1}7s^{2}$<br>Protactinium | **U** 92<br>238.03<br>1.38 6,5,4,3<br>— 1.7<br>$[Rn]5f^{3}6d^{1}7s^{2}$<br>Uranium | **Np** 93<br>(237)<br>1.3 6,5,4,3<br>— 1.3<br>$[Rn]5f^{4}6d^{1}7s^{2}$<br>Neptunium | **Pu** 94<br>(242)<br>1.51 6,5,4,3<br>— 1.3<br>$[Rn]5f^{6}6d^{0}7s^{2}$<br>Plutonium | **Am** 95<br>(243)<br>1.73 6,5,4,3<br>— 1.3<br>$[Rn]5f^{7}6d^{1}7s^{2}$<br>Americium |

Nonmetals

Metals

# THE ELEMENTS

| | | | | | | | | | | | | | | | | | | | | |
|---|---|---|---|---|---|---|---|---|---|---|---|---|---|---|---|---|---|---|---|---|
| .2 | 1.3 | 1.4 | 1.5 | 1.6 | 1.7 | 1.8 | 1.9 | 2.0 | 2.1 | 2.2 | 2.3 | 2.4 | 2.5 | 2.6 | 2.7 | 2.8 | 2.9 | 3.0 | 3.1 | 3.2 |
| 0 | 34 | 39 | 43 | 47 | 51 | 55 | 59 | 63 | 67 | 70 | 74 | 76 | 79 | 82 | 84 | 86 | 88 | 89 | 91 | 92 |

Legend (97):
- ATOMIC NUMBER
- ATOMIC MASS (2)
- OXIDATION STATES (Bold most stable)
- ELECTRONEGATIVITY (Pauling's)
- ELECTRON STRUCTURE

Groups: IB, IIB, IIIA, IVA, VA, VIA, VIIA, VIIIA

| Symbol | Atomic number | Atomic mass | | | Oxidation states | Electronegativity | Electron structure | Name |
|---|---|---|---|---|---|---|---|---|
| He | 2 | 4.0026 | 0.32 | | — | — | $1s^2$ | Helium |
| B | 5 | 10.811 | 0.98 | 191 | 3 | 2.0 | $1s^2 2s^2 2p^1$ | Boron |
| C | 6 | 12.01115 | 0.914 | 260 | 4,2 | 2.5 | $1s^2 2s^2 2p^2$ | Carbon |
| N | 7 | 14.0067 | 0.92 | 336 | 3,6,4,2 | 3.0 | $1s^2 2s^2 2p^3$ | Nitrogen |
| O | 8 | 15.9994 | 0.73 | 314 | -2 | 3.5 | $1s^2 2s^2 2p^4$ | Oxygen |
| F | 9 | 18.9984 | 0.71 | 402 | -1 | 4.0 | $1s^2 2s^2 2p^5$ | Fluorine |
| Ne | 10 | 20.183 | 0.69 | 497 | — | — | $1s^2 2s^2 2p^6$ | Neon |
| Al | 13 | 26.9815 | 1.43 | 138 | 3 | 1.5 | $[Ne]3s^2 3p^1$ | Aluminum |
| Si | 14 | 28.086 | 1.17 | 188 | 4 | 1.8 | $[Ne]3s^2 3p^2$ | Silicon |
| P | 15 | 30.9738 | 1.10 | 254 | 3,5,4 | 2.1 | $[Ne]3s^2 3p^3$ | Phosphorus |
| S | 16 | 32.064 | 1.04 | 239 | 2,4,6 | 2.5 | $[Ne]3s^2 3p^4$ | Sulfur |
| Cl | 17 | 35.453 | 0.99 | 300 | 1,3,5,7 | 3.0 | $[Ne]3s^2 3p^5$ | Chlorine |
| Ar | 18 | 39.948 | 0.97 | 363 | — | — | $[Ne]3s^2 3p^6$ | Argon |
| Ni | 28 | 58.71 | 1.24 | 176 | 2,3 | 1.8 | $[Ar]3d^8 4s^2$ | Nickel |
| Cu | 29 | 63.54 | 1.28 | 178 | 2,1 | 1.9 | $[Ar]3d^{10} 4s^1$ | Copper |
| Zn | 30 | 65.37 | 1.38 | 216 | 2 | 1.6 | $[Ar]3d^{10} 4s^2$ | Zinc |
| Ga | 31 | 69.72 | 1.22 | 138 | 3 | 1.6 | $[Ar]3d^{10} 4s^2 4p^1$ | Gallium |
| Ge | 32 | 72.59 | 1.22 | 187 | 4 | 1.8 | $[Ar]3d^{10} 4s^2 4p^2$ | Germanium |
| As | 33 | 74.922 | 1.21 | 231 | 3,5 | 2.0 | $[Ar]3d^{10} 4s^2 4p^3$ | Arsenic |
| Se | 34 | 78.96 | 1.17 | 225 | -2,4,6 | 2.4 | $[Ar]3d^{10} 4s^2 4p^4$ | Selenium |
| Br | 35 | 79.909 | 1.14 | 273 | 1,5 | 2.8 | $[Ar]3d^{10} 4s^2 4p^5$ | Bromine |
| Kr | 36 | 83.80 | 1.10 | 323 | — | — | $[Ar]3d^{10} 4s^2 4p^6$ | Krypton |
| Pd | 46 | 106.4 | 1.37 | 192 | 2,4 | 2.2 | $[Kr]4d^{10} 5s^0$ | Palladium |
| Ag | 47 | 107.870 | 1.44 | 175 | 1 | 1.9 | $[Kr]4d^{10} 5s^1$ | Silver |
| Cd | 48 | 112.40 | 1.54 | 207 | 2 | 1.7 | $[Kr]4d^{10} 5s^2$ | Cadmium |
| In | 49 | 114.82 | 1.66 | 133 | 3 | 1.7 | $[Kr]4d^{10} 5s^2 5p^1$ | Indium |
| Sn | 50 | 18.69 | 1.40 | 169 | 4,2 | 1.8 | $[Kr]4d^{10} 5s^2 5p^2$ | Tin |
| Sb | 51 | 121.75 | 1.41 | 199 | 3,5 | 1.9 | $[Kr]4d^{10} 5s^2 5p^3$ | Antimony |
| Te | 52 | 127.60 | 1.43 | 208 | -2,4,6 | 2.1 | $[Kr]4d^{10} 5s^2 5p^4$ | Tellurium |
| I | 53 | 126.904 | 1.33 | 241 | 1,5,7 | 2.5 | $[Kr]4d^{10} 5s^2 5p^5$ | Iodine |
| Xe | 54 | 131.30 | 1.30 | 280 | — | — | $[Kr]4d^{10} 5s^2 5p^6$ | Xenon |
| Pt | 78 | 195.09 | 1.39 | 207 | 2,4 | 2.2 | $[Xe]4f^{14} 5d^9 6s^1$ | Platinum |
| Au | 79 | 196.967 | 1.46 | 213 | 3,1 | 2.4 | $[Xe]4f^{14} 5d^{10} 6s^1$ | Gold |
| Hg | 80 | 200.59 | 1.57 | 241 | 2,1 | 1.9 | $[Xe]4f^{14} 5d^{10} 6s^2$ | Mercury |
| Tl | 81 | 204.37 | 1.71 | 141 | 3,1 | 1.8 | $[Xe]4f^{14} 5d^{10} 6s^2 6p^1$ | Thallium |
| Pb | 82 | 207.19 | 1.75 | 171 | 4,2 | 1.8 | $[Xe]4f^{14} 5d^{10} 6s^2 6p^2$ | Lead |
| Bi | 83 | 208.980 | 1.70 | 185 | 3,5 | 1.9 | $[Xe]4f^{14} 5d^{10} 6s^2 6p^3$ | Bismuth |
| Po | 84 | (210) | 1.76 | — | 2,4 | 2.0 | $[Xe]4f^{14} 5d^{10} 6s^2 6p^4$ | Polonium |
| At | 85 | (210) | — | — | 1,3,5,7 | 2.2 | $[Xe]4f^{14} 5d^{10} 6s^2 6p^5$ | Astatine |
| Rn | 86 | (222) | — | 248 | — | — | $[Xe]4f^{14} 5d^{10} 6s^2 6p^6$ | Radon |

| Symbol | Atomic number | Atomic mass | | | Oxidation states | Electronegativity | Electron structure | Name |
|---|---|---|---|---|---|---|---|---|
| Gd | 64 | 157.25 | 1.79 | 142 | 3 | 1.1 | $[Xe]4f^7 5d^1 6s^2$ | Gadolinium |
| Tb | 65 | 158.924 | 1.80 | 155 | 3,4 | 1.2 | $[Xe]4f^9 5d^0 6s^2$ | Terbium |
| Dy | 66 | 162.50 | 1.80 | 157 | 3 | — | $[Xe]4f^{10} 5d^0 6s^2$ | Dysprosium |
| Ho | 67 | 164.930 | 1.79 | — | 3 | 159 | $[Xe]4f^{11} 5d^0 6s^2$ | Holmium |
| Er | 68 | 167.26 | 1.78 | — | 3 | 1.2 | $[Xe]4f^{12} 5d^0 6s^2$ | Erbium |
| Tm | 69 | 168.934 | 1.77 | — | 3,2 | 1.2 | $[Xe]4f^{13} 5d^0 6s^2$ | Thulium |
| Yb | 70 | 173.04 | 1.94 | 143 | 3,2 | 1.1 | $[Xe]4f^{14} 5d^0 6s^2$ | Yitterbium |
| Lu | 71 | 174.97 | 1.75 | 115 | 3 | 1.2 | $[Xe]4f^{14} 5d^1 6s^2$ | Lutetium |
| Cm | 96 | (247) | — | — | 3 | — | $[Rn]5f^7 6d^1 7s^2$ | Curium |
| Bk | 97 | (247) | — | — | 4,3 | — | $[Rn]5f^9 6d^0 7s^2$ | Berkelium |
| Cf | 98 | (249) | — | — | 3 | — | $[Rn]5f^{10} 6d^0 7s^2$ | Californium |
| Es | 99 | (254) | — | — | — | — | $[Rn]5f^{11} 6d^0 7s^2$ | Einsteinium |
| Fm | 100 | (253) | — | — | — | — | $[Rn]5f^{12} 6d^0 7s^2$ | Fermium |
| Md | 101 | (256) | — | — | — | — | $[Rn]5f^{13} 6d^0 7s^2$ | Mendelevium |
| No | 102 | (254) | — | — | — | — | $[Rn]5f^{14} 6d^0 7s^2$ | Nobelium |
| Lw | 103 | (257) | — | — | — | — | $[Rn]5f^{14} 6d^1 7s^2$ | Lawrencium |

# GLOSSARY

**adept** . . . . . . . . . . . . . . . . . . . . . . . . . . . . . An expert in the science of alchemy. One who has been divinely given the true secrets of alchemy. A master from whom others were to learn.

**alchemy** . . . . . . . . . . . . . . . . . . . . . . . . . . An ancient study or art which attempted to combine science and philosophy to solve worldly problems.

**alloy** . . . . . . . . . . . . . . . . . . . . . . . . . . . . . A homogeneous mixture of elements with metals to form metallic solids.

**atom** . . . . . . . . . . . . . . . . . . . . . . . . . . . . . The simplest unit of an element that still retains the properties of that element.

**chemical change** . . . . . . . . . . . . . . . . . . . A change that causes the starting materials to completely lose their properties to form new and distinctly different substances.

**colloid** . . . . . . . . . . . . . . . . . . . . . . . . . . . A uniform dispersion, heterogeneous in nature between two or more substances.

**compound** . . . . . . . . . . . . . . . . . . . . . . . . Two or more elements joined together such that the elements have lost their individual identity in favor of a new set of properties.

**element** . . . . . . . . . . . . . . . . . . . . . . . . . . A primary substance that cannot be divided into separate substances; one of about 106 different basic varieties of matter making up the universe.

**heterogeneous** . . . . . . . . . . . . . . . . . . . . Composed of dissimilar parts which can be separated easily and which are unevenly distributed in the mixture.

**homogeneous** . . . . . . . . . . . . . . . . . . . . . Even distribution of parts throughout the whole mixture; not easily separated into individual components.

**inorganic** . . . . . . . . . . . . . . . . . . . . . . . . . From nonliving material.

**metal** . . . . . . . . . . . . . . . . . . . . . . . . . . . . The group of elements that are good conductors of electricity and heat; solids except for mercury.

**mixture** . . . . . . . . . . . . . . . . . . . . . . . . . . Two or more substances dispersed in one another but each retaining their own identity.

**nonmetal** . . . . . . . . . . . . . . . . . . . . . . . . The group of elements that are poor conductors of electricity and heat; mostly liquids and gases.

**organic** . . . . . . . . . . . . . . . . . . . . . . . . . . Derived from a living or a once-living substance.

**phase change** . . . . . . . . . . . . . . . . . . . . . A change from one state of matter to another.

**physical edge** . . . . . . . . . . . . . . . . . . . . . A change in shape, size, or physical characteristics but remaining the same substance.

**puffer** . . . . . . . . . . . . . . . . . . . . . . . . . . . . An alchemist who concentrated solely on turning base metals into gold. They were mocked by the true alchemists or adepts for their use of bellows in the process of heating base metals.

**solution** . . . . . . . . . . . . . . . . . . . . . . . . . . A homogeneous mixture, usually a liquid with some substance dissolved in it.

**suspension** . . . . . . . . . . . . . . . . . . . . . . . A temporary heterogeneous mixture that separates into individual components with time.